The Application of VARIABLE SPEED DRIVES

David William Spitzer

INSTRUMENT SOCIETY OF AMERICA

The Application of
VARIABLE SPEED DRIVES

Printed in the United States of America

INSTRUMENT SOCIETY OF AMERICA
67 Alexander Drive
P.O. Box 12277
Research Triangle Park, NC 27709

Library of Congress Cataloging-in-Publication Data

Spitzer, David W.
The application of variable speed drives.

Bibliography: p.
1. Variable speed drives. I. Title.
TJ1051.S75 1987 621.8′12 87-3585
ISBN 0-87664-955-X

Book design by Summit Technical Associates, Inc.

Dedication

This work is dedicated to my children, Deborah Goldie and Michael James, in the hope that it will inspire them and others to continue their education and their pursuit of knowledge; and to my wife Ida for understanding why I was so late to bed for so many months.

Acknowledgments

My goal in writing this text is to help those in industry to understand, evaluate, and exploit a technology that has the potential to significantly reduce the operating costs associated with virtually any piece of rotating equipment. I would like to thank T. C. Haines and L. H. Chang for their help in reviewing the manuscript.

Table of Contents

Introduction

1

Variable speed drive technology spans virtually the entire industrial revolution, from the development of the steam engine to modern microprocessor-based electronic motor controllers. Applications of the technology range from miniature motors that can be held in one's hand to pieces of rotating equipment that are considerably larger than the tallest man. Refinements of this technology, fueled in recent years by improved reliability and performance of electronic devices, are occurring at a more rapid pace than ever, resulting in greater performance at lower cost. The net result is a new world of technical and economic opportunity for those who are astute enough to grasp the potential of this technology in its various and sundry forms.

Simply stated, variable speed drive technology encompasses those devices that alter the speed of rotating equipment. The ramifications of the above statement manifest themselves in many disciplines within the engineering profession, which, perhaps to a great degree, accounts for some lack of exploitation of this technology.

The start of the design cycle typically involves a *process* engineer who is an expert in one area and who may have a cursory knowledge of related disciplines; thus, the initial concept is developed with strong emphasis on the technical correctness and viability of the process, but with virtually no emphasis on its final implementation. Therefore, the decision to apply variable speed drive technology is almost always left to others during this stage of project development.

Once development is complete, the *project* engineer is given the task of implementing the project in a technically correct manner within time and budgetary constraints. The project engineer (who may be a chemical, civil, electrical, instrumentation, manufacturing, mechanical, or plant engineer) may not have sufficient knowledge of variable speed drive technology to consider its application.

Budgets, completion dates, and many other project-related problems tend to take precedence over investigating the application of this technology, especially since more conventional technology and a more understandable approach may appear to require less time and resources to implement, perhaps at the expense of operational results. Many good project engineers suffer from the "get the best delivery" and the "buy the cheapest one because they are all the same" syndromes and have lost sight of installing the *best possible system*, given the constraints of technical correctness, time, and budget. It is, therefore, not uncommon for a project engineer to determine the applicability of variable speed drive technology through expediency rather than on a technical basis.

In its most prevalent form, variable speed drive technology manifests itself as electronic switchgear, seemingly within the realm of the *electrical* engineer. Enter the *instrumentation* engineer, who feels that a variable speed drive is a final control element, and a conflict can arise as the instrumentation engineer may feel that this technology falls within the instrument domain. The individuals in both of these disciplines do, however, share common bonds and frustrations: their "gut" feelings about the applicability of variable speed drive technology; having to prove their point with only limited expertise in other engineering disciplines; arriving late in the project (which is a time when people who propose design changes are frowned upon); and dealing with a project engineer who typically does not understand and does not *want* to understand why what was never a problem before is a problem now.

Given the above scenario, it can be seen that many applications of variable speed drive technology have fallen between the cracks of the traditional engineering disciplines, resulting in lost opportunities and ignoring a great number of potential retrofit applications in existing plants and processes.

In virtually all organizations, management will control expenditures and manpower of projects that utilize variable speed drive technology. It must be accepted that the majority of managers who have ultimate control over these projects either never were or are no longer technically oriented. A successful project must be documented in a technically sound manner and translated into language that management knows, understands and is interested in — in other words, the language of economic impact. Projects that are proposed on the basis of what *should* be done are understood to be "wish list" items and can be readily ignored; those that are vague "lack engineering substance"; "qualitative results" are not believable and results with specific technical ramifications are generally not understood and therefore cannot be acted upon; while those projects that *must* be done are bothersome problem areas. On the other hand, projects that have a short payback period are readily understood and relished by management because they make the individual who proposed them and the manager who implemented them look good.

A detailed engineering and economic analysis is one straightforward method used to convince management that a project is justified. Letting management *see* the calculations and the logic involved is one of the best methods for convincing management to implement a project.

The intent of this book is to give the reader the tools to perform a technically sound evaluation of a concept involving the application of variable speed drive technology. This approach is multi-discipline in nature, involving combinations of the fields of architecture, chemistry, chemical engineering, electrical engineering, hydraulics, instrumentation, mechanical engineering, physics, piping, and utilities, to name a few. These topics are covered only to the degree necessary to develop a technically sound evaluation for most applications. However, the underlying concepts used to perform an evaluation can be coupled with the topics covered in the text and the reader's knowledge of the process at hand to perform a valid technical evaluation of a potential application.

It must be understood that a technical evaluation is nothing more than an estimate of the expected system performance. Various uncertainties, including operating conditions that are different from design conditions, can affect actual system operation. Therefore, in many cases, rules of thumb and estimating techniques are used to describe complex phenomena in an effort to avoid minute details and lengthy analyses, thereby obtaining approximate results in a more efficient and expedient manner.

Real-world examples and calculations are presented throughout the text to illustrate the application of the principles. As presented herein, the gauge of the economic evaluation of a project is limited to its simple payback, with the reader being left to investigate the exact method utilized within the reader's organization.

In summary, variable speed drive technology has often been overlooked and ignored by the traditional engineering disciplines due to the multi-disciplined background necessary to properly evaluate most applications. This book is an attempt to bridge the gaps in the reader's technical background so that the reader can apply variable speed drive technology where it is justified.

Electrical Basics

2

Unless one is designing or maintaining electronic equipment, everyday practical information becomes much more important than the laws of electricity that are presented in the classical texts. The intent of this chapter is to present those electrical concepts in a practical manner consistent with the understanding of the ensuing analysis. While some of the knowledge may seem trivial to some readers, the assumptions and models presented herein will be used throughout the text.

VOLTAGE AND CURRENT

Classically, there exists a potential difference between two points in an electrostatic field. Electrical work is performed when a charge is moved between these two points. All things being equal, the higher the potential difference, the more charge will be moved and the more electrical work will be performed.

The potential difference between the two points is termed the *voltage* and is measured in volts (V), while the movement of charge is termed the *current* and is measured in amperes (A). The hydraulic analogies for voltage and current are pressure and flow, respectively. Electrical resistance opposes the flow of current and is hydraulically equivalent to a restriction to flow.

The relationship between voltage and current in a simple resistive circuit is given by Ohm's Law

$$V = I \times R$$

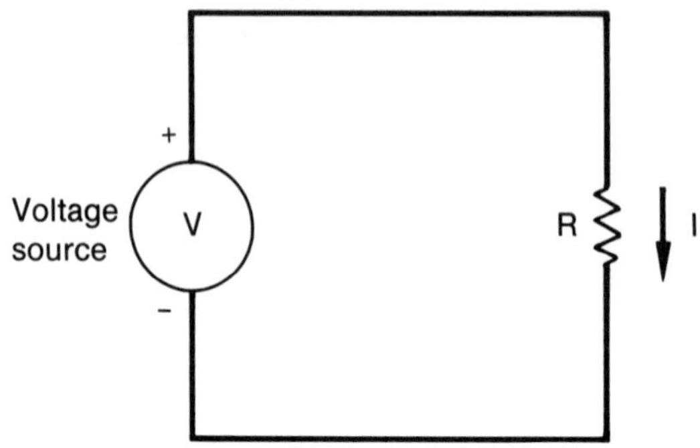

Figure 2.1
Simple Resistive Circuit

where V is the potential difference in volts, I is the current in amperes and R is the resistance in ohms (see Figure 2.1). The voltage and current are termed dc (for direct current) when the voltage is constant, thereby causing a constant current to flow (see Figure 2.2).

The voltage and current are termed ac (for alternating current) when the voltage changes with time to form a sine wave, thereby causing a current to flow in the form of a sine wave (see Figure 2.3).

Countless other voltage waveforms can be generated to produce different current waveforms; however, ac and dc voltage waveforms predominate industrial power distribution.

A battery is a device that generates a dc waveform, as the voltage is held constant with time during its useful life. An example of an ac voltage source is a wall outlet, at which, depending upon location, the voltage oscillates at 50 or 60 hertz (cycles per second) in the shape of a sine wave.

Both ac and dc power sources possess the potential to do work when connected to a proper load. Power consumption is given by

$$P = I \times V$$

where P is the power in watts (W). Combining the above equation with Ohm's Law yields

$$P = I^2 \times R$$

from which the power consumed by the resistance can be calculated. Due to the multitude of components that comprise a device, real circuits contain not only resistance, but capacitance and inductance as well.

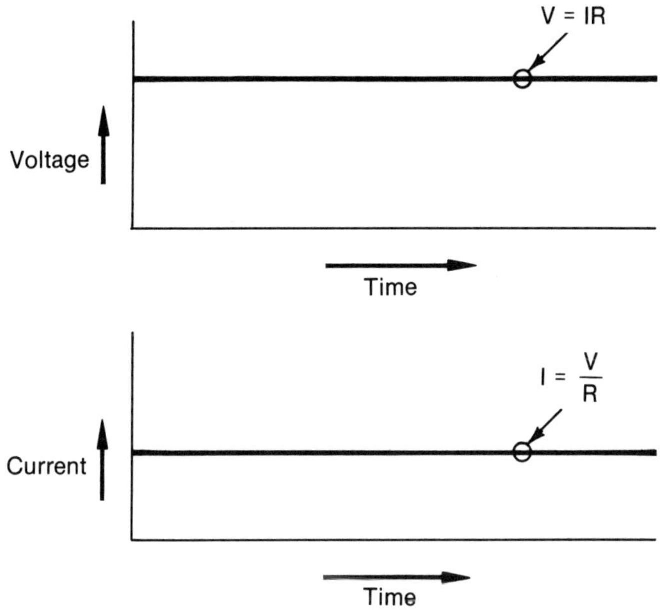

Figure 2.2
DC Excitation

Capacitance is the ability of the device to store energy in the form of an electric charge. Due to the nature of the capacitor (typically, two parallel plates closely separated by an insulator), when a voltage is applied and abruptly removed, perhaps by disconnecting the capacitor from the circuit, the voltage across the capacitor is retained. Connecting the capacitor to a load will cause current to flow, discharging the capacitor and eventually reducing the voltage across the capacitor to zero.

Inductance is the ability of a device to store energy in the form of a magnetic field. When a constant voltage is applied and abruptly removed from an inductor (typically, an electromagnetic coil) the inductor uses its stored magnetic energy to maintain a constant current through the inductor. Under these circumstances (that is, when the inductor attempts to maintain constant current and the resistance of the circuit is increased abruptly), according to Ohm's Law, the voltage across the inductor will be very large until the energy stored in the inductor is dissipated. Therefore, attempting to disconnect operating inductive devices, such as motors or current transformers, can induce high voltage spikes that can result in physical injury.

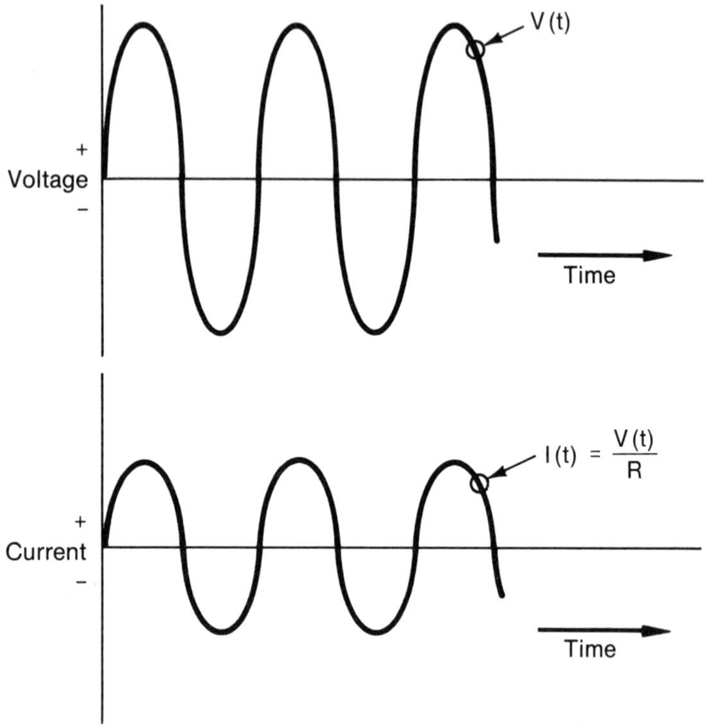

Figure 2.3
AC Excitation

In circuits with resistance, capacitance, and inductance, the relationship between voltage, current, and load is given by

$$V = I \times Z$$

where Z is the impedance of the circuit, which takes into account the combined effects of resistance, capacitance, and inductance. Combining this equation with that for power yields

$$P = I^2 \times Z$$

The *impedance* is a complex number and is the sum of its so-called real and imaginary parts. Similarly, in this case, the power consists of both real and imaginary components.

The real part of the impedance is the overall resistance of the circuit, which can be used to calculate the real power that a device consumes. The real power is measured in watts (W) and can be thought of as the power that is used to do useful work and generate heat due to inefficiencies in a device.

The imaginary part of the impedance describes the storage effects of capacitors and inductors and can be used to calculate the imaginary power that the device consumes. The imaginary power is measured in volt-amperes reactive (VAR) and can be thought of as the power that is not used to do useful work or generate heat but is used to charge capacitors and/or magnetize inductors.

The concepts of real and imaginary parts of power are important in that current is necessary to provide not only the real power but also the imaginary power. The amount of current required to operate the load is measured in volt-amperes (VA) and is determined by performing a vector addition of the real and imaginary currents as illustrated in Figure 2.4, where from trigonometry,

$$I_{real} = I \times \cos\theta$$

where θ is the angle formed by I and I_{real}. As θ varies from 0 to 90 degrees, the cosine function has values between 1 and 0, respectively. This means that the current that actually flows through the device is greater than or equal to the current required to perform real work. The real power delivered to a device is the instantaneous power, which is the product of the instantaneous voltage and the instantaneous current, given by

$$P = V \times I \times \cos\theta$$

Keeping in mind that a device will draw the amount of real power it needs to operate, θ represents a *phase shift* between the voltage and current.

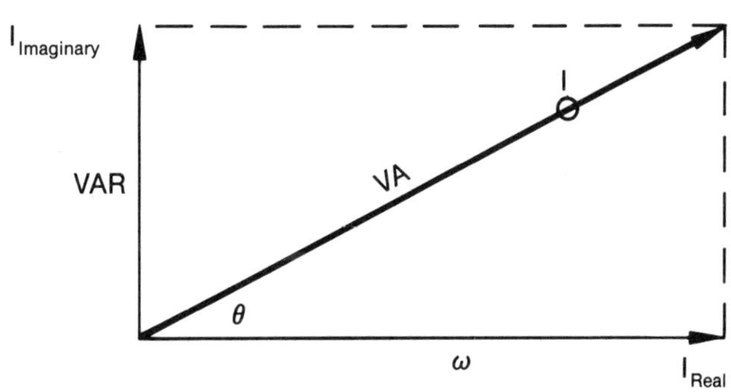

Figure 2.4
Vector Addition of Real and Imaginary Current

The phase shift occurs due to the effects of capacitance and inductance, which, in loose terms, tend to store voltage and current, respectively.

The term cos θ in the above equation represents the factor by which the actual current exceeds the real part of the current and is commonly called the *power factor*.

Power Factor (PF) = cos θ

From the above equations, it can be seen that the power factor is inversely related to the amount of current drawn in a circuit. The real power (watts) used to satisfy a given load requirement will remain virtually constant, independent of the power factor. The power factor of the circuit can be thought of as the ratio of the current required to provide the same power to a purely resistive load (unity power factor) to the actual current.

As an example, a load with a power factor of 0.4 will require twice as much current as the same load with a power factor of 0.8, and 2.5 times as much current as the same resistive load.

SINGLE-PHASE AND THREE-PHASE POWER

A load is termed *single-phase* when it is operated by one voltage source. Common electrical receptacles are typical single-phase power sources (see Figure 2.5).

A load is termed *three-phase* when it is operated by three voltage sources whose waveforms are 120 degrees out of phase with one another (see Figure 2.6). Utility power and industrial power are distributed in this manner, and

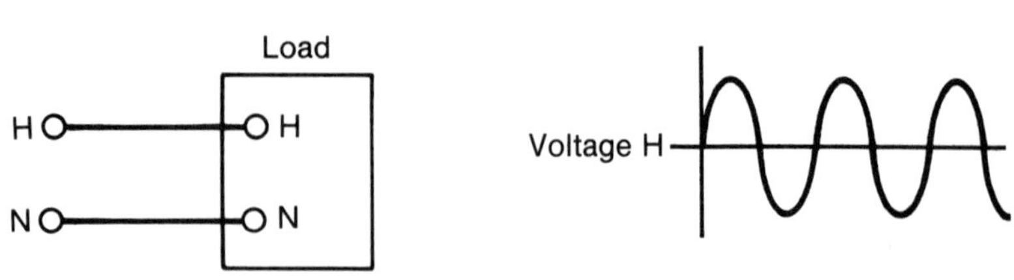

Figure 2.5
Single-Phase Load

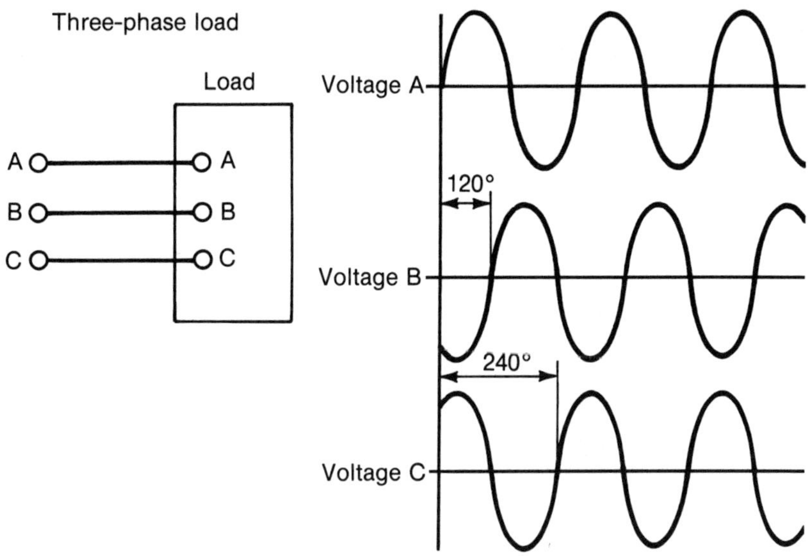

Figure 2.6
Three-Phase Load

all but the smallest industrial loads are designed to operate using three-phase power.

The equations for single-phase and three-phase power, respectively, can be calculated in kilowatts (kW) by the following useful relationships.

$$kW = \frac{V \times A \times PF}{1000} \qquad \text{(Single-phase)}$$

$$kW = \frac{V \times A \times PF \times 1.732}{1000} \qquad \text{(Three-phase)}$$

TRANSFORMERS

Power companies generate and transmit voltages that have electrical potentials measuring thousands of volts. *Transformers* are used to increase these voltages for more efficient long distance transmission or to reduce voltages for compatibility with the user's electrical system.

Conceptually, the transformer can be thought of as a magnetic isolator between two circuits. Physically, the primary and secondary coils are wound on a common ferrous core. When the primary coil is excited by an ac waveform, the core becomes magnetized and in turn induces a voltage and current in the secondary coil. Assuming that the transformer has negligible losses associated with its operation, the relationship between the number of turns in the windings, voltages and currents is predictable (Figure 2.7).

The ratio of the secondary voltage to the primary voltage is fixed so that, given a constant primary voltage, the secondary voltage will be constant. Under constant voltage conditions, the ratio of the primary current to the secondary current will also be constant. Therefore, an increase (decrease) of secondary current will increase (decrease) the primary current by the same percentage. Transformers are typically rated in kilovolt-amperes (kVA), which reflects the current handling capacity of the transformer, where

$$\text{kVA} = \frac{\text{V} \times \text{A}}{1000} \qquad \text{(Single-phase)}$$

$$\text{kVA} = \frac{\text{V} \times \text{A} \times 1.732}{1000} \qquad \text{(Three-phase)}$$

Real transformers exhibit losses of approximately 0.5 percent of the full load of the transformer plus 1.5 percent of the actual transformer kVA load. The losses are percentages of the power that would be produced by the load with unity power factor; however, the actual transformer loading is a function of kVA (as opposed to power).

For example, a three-phase 1000 kilovolt-ampere (kVA) transformer, which can provide 1000 kilowatts (kW) to a load with a power factor of 1.0, operating at 400 kVA would have losses of approximately [(0.005 × 1000 kW) + 0.015 × (400/1000) × 1000 kW] or 11 kW.

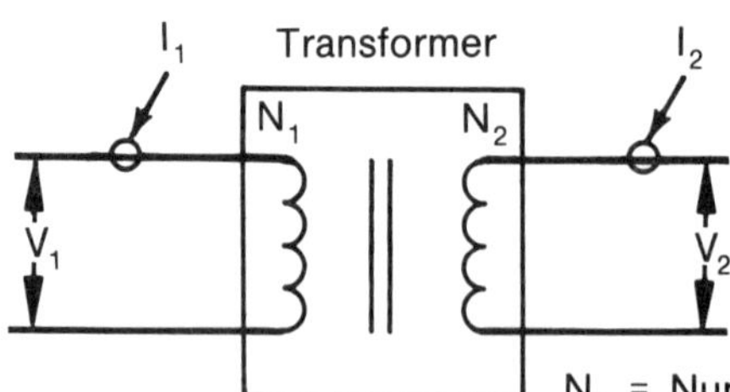

$$\frac{V_1}{V_2} = \frac{N_1}{N_2} = \frac{I_2}{I_1}$$

N_1 = Number of turns in primary coil

N_2 = Number of turns in secondary coil

Figure 2.7
Model of a Transformer

MOTORS

Electric motors are the most prevalent prime movers used to operate rotating equipment. There are many designs of motors and variations thereof, but ac induction motors and dc motors are most commonly applied in industry. Complex equations and models can be used to describe the operation of these devices; in this text simple explanations and models will be used to approximately model their performance.

The capacity (size) of a motor is the amount of mechanical work the motor will perform. Motor capacities are stated in kilowatts (kW) or horsepower (hp), where

$$1 \text{ hp} = 0.746 \text{ kW}$$

The amount of electrical energy that must be fed to the motor will be greater than the amount of useful energy that is transmitted to the motor shaft (load) due to the various loss mechanisms present in the motor. The motor efficiency, which is the ratio of output energy to input energy, is given by

$$\frac{\text{mechanical power output}}{\text{electrical power input}}$$

which is a function of the motor and its operating conditions, for example, load and speed.

AC Induction Motors

AC induction motors, often called squirrel-cage motors, are applied extensively in industry due to their rugged design, durability, and low maintenance requirements. These motors can be manufactured for use in hazardous locations and do not contain any brushes or other sparking devices.

A fully loaded induction motor will have relatively high efficiency and a high power factor. As the loading drops, both the efficiency and the power factor drop significantly. An unloaded induction motor will draw approximately 35 percent of its full load current and will consume approximately 15–30 percent of its full load power consumption, depending upon size and construction.

In order to model the above phenomenon in a simple manner, the following approximations are made. Power consumption at no load conditions is estimated to be 15–33 percent of full load power consumption, depending on motor size and construction, where larger motors consume a smaller percentage of full load power. Full load power consumption is consumed at full load conditions, and power consumption varies linearly between these two load conditions. These assumptions form a basis with which power consumption can be estimated as a function of full load power consumption when a motor is operated at various load conditions.

Under full load operating conditions, motor efficiency is approximately 75–93 percent, depending on motor size and construction, where larger motors exhibit higher efficiencies. This suggests an approximate relationship between motor capacity and the input power when the motor is fully loaded.

Combining the above assumptions, the relationship between power consumption of a partially loaded motor and motor capacity can be approximated mathematically as

$$\text{Power Consumption} = \left[\frac{(0.15 + \Delta\,\eta) + (0.85 - \Delta\,\eta)\left(\dfrac{\text{Actual Load}}{\text{Motor Capacity}}\right)}{(0.93 - \Delta\,\eta)}\right](\text{Motor Capacity})$$

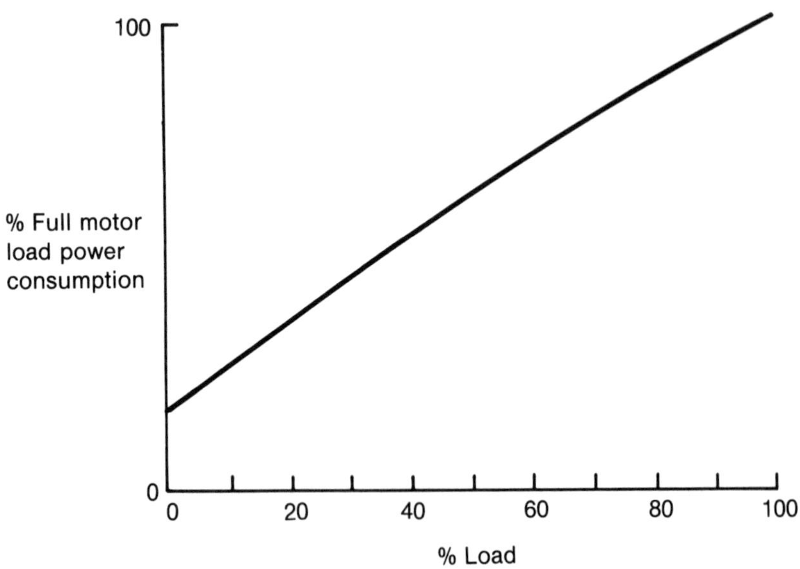

Figure 2.8
Model of Power Consumption vs. Load of an AC Induction Motor

where $\Delta\eta$ is given by:

Table 2.1

Motor Capacity		$\Delta\eta$
0–3 hp	0–2.2 kW	0.18
5–10	3.7–7.5	0.15
15–25	11.2–18.7	0.12
30–40	22.4–44.8	0.09
50–60	37.3–44.8	0.06
75–100	56.0–74.6	0.03
over 125	over 93.3	0.00

As an example, the input power requirement of a 100 hp (74.6 kW) motor under a 40 kW mechanical load can be estimated as

$$\left[(0.15 + 0.03) + (0.85 - 0.03) \times \left(\frac{40 \text{ kW}}{74.6 \text{ kW}}\right)\right] \times \frac{74.6 \text{ kW}}{(0.93 - 0.03)} = 51.4 \text{ kW}$$

Note that the motor efficiency at this load condition can be estimated as (40 kW/51.4 kW), or 78 percent, which represents a significant reduction from the 90 percent estimated efficiency of a fully loaded motor.

Note that, for smaller motors, motor efficiency percentage approximates the relationship between kilowatts and horsepower, such that the fully loaded power consumption (kW) is approximately numerically equal to the horsepower rating of the motor. This approximation should be used only for preliminary calculations for motors under 20 hp (14.9 kW).

Similarly, the amount of current as a percentage of full load amperes (FLA) that an induction motor will draw can be roughly estimated as

$$\text{Current} = \left[0.35 + 0.65\left(\frac{\text{Actual Load}}{\text{Motor Capacity}}\right)\right] (\text{FLA})$$

DC Motors

DC motors have been used for many years in many applications in sizes from fractional through thousands of horsepower. Armature excitation as well as field excitation is required to operate a dc motor. Brushes are used to change the direction of the armature current as the motor rotates. The arcing of the brushes limits the suitability of dc motors in hazardous applications, while brush wear introduces maintenance problems.

Figure 2.9
DC Motor
(courtesy Allen Bradley)

Hydraulic Basics

3

Hydraulics play an extremely important role in industrial systems and must be dealt with during the design stage as well as in virtually all process modifications. Predicting an approximate hydraulic response of the process is often essential to successful system operation. The hydraulic properties of some commonly used equipment will be discussed to develop a working knowledge of their operating characteristics.

PUMPS

Pumps, typically driven by an electric motor or steam turbine, are used throughout industry to develop hydraulic energy. Liquid enters the pump suction (inlet) where the rotation of the pump imparts pressure energy to the liquid that exits at the pump discharge (outlet).

The pressure at the inlet (outlet) of the pump is termed the *suction (discharge) pressure* or *head*. The amount of pressure energy that is gained by the liquid is called the *total dynamic head* (TDH) of the pump and is the pressure difference between the pump suction and discharge pressures. TDH is typically expressed in meters (feet) of water column and is independent of the specific gravity of the liquid being pumped.

The amount of hydraulic energy imparted to the liquid can be calculated using the following formulas, where SG is the specific gravity of the liquid.

$$\text{Pump hydraulic output} = (Q_{\text{gpm}} \times \text{TDH}_{\text{ft}} \times \text{SG})/3960 \text{ hp}$$
$$= (Q_{\text{lpm}} \times \text{TDH}_{\text{m}} \times \text{SG})/4601 \text{ kW}$$

Because pumping applications and operating conditions vary from process to process, many different pump designs have evolved, such as vertical, submersible, metering, diaphragm, positive displacement, centrifugal, and the like, each with its unique performance characteristics.

The most commonly applied pump design is the *centrifugal pump*, where liquid enters the pump at the center of the impeller, the rotation of which causes the liquid to be moved outward (centrifugally) to the pump discharge, thereby imparting hydraulic energy to the liquid (see Figure 3.1).

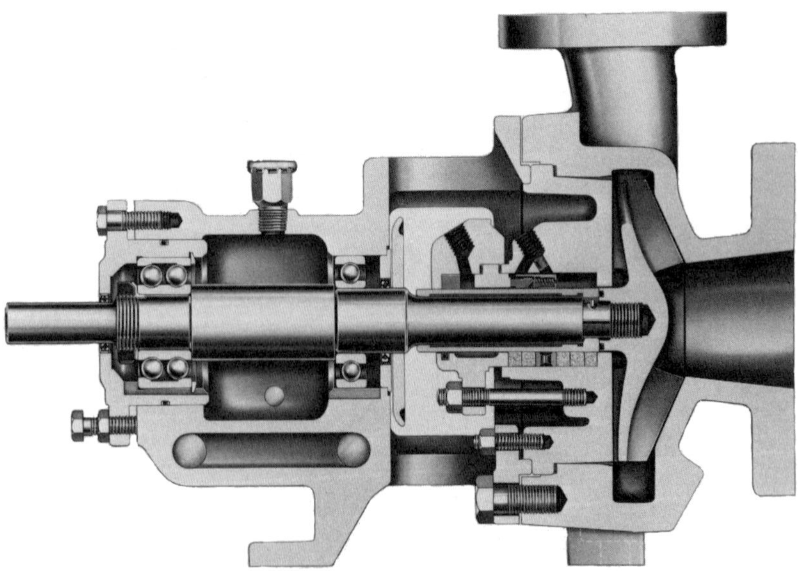

Figure 3.1
Typical Centrifugal Pump
(courtesy Gould Pumps)

The *positive displacement pump*, which entraps liquid at its suction and transfers the liquid to the discharge (causing a flowing condition), is also widely applied (see Figure 3.2).

PUMP PERFORMANCE CURVES

Pump performance is usually described graphically on what are commonly called the *pump curves* published in pump manufacturers' literature. One graph often contains virtually all of the pump operating data and can be confusing unless properly understood and interpreted.

Due to the wide applicability, reliability, and general user acceptance of centrifugal pumps, their performance is used to illustrate typical techniques by which pump performance is most commonly expressed. The shape of the pump curves for specific centrifugal and other pump designs may be very different from those presented for the typical centrifugal pump. Manufacturer documentation should be consulted to determine the effects of increasing liquid viscosity on pump performance (when applicable).

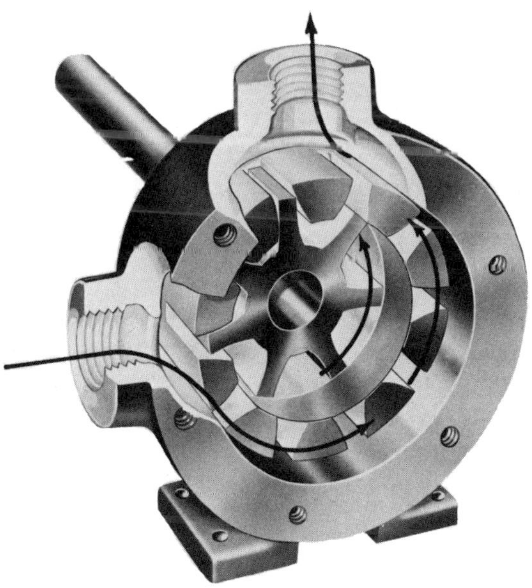

Figure 3.2
Typical Positive Displacement Pump
(courtesy Viking Pump-Houdaille™)

Head

The TDH that the pump will generate varies with the flow through the pump and is commonly called the *head* (see Figure 3.3a).

From the graph in Figure 3.3a, the pump head can be determined at each flow rate. It should be noted that the head of a centrifugal pump is typically at its maximum when the pump is under no flow conditions (dead-headed) and decreases with increasing flow.

Typical manufacturer pump curves depict the head characteristics of different impeller sizes on the same graph. Note that as impeller size increases, the pump is capable of producing more flow and pressure, as shown in Figure 3.3b.

Brake Horsepower

The *brake horsepower* is the amount of rotational energy required by the pump at its shaft under given operating conditions. Brake horsepower at extreme operating conditions is one criterion used to determine motor size.

Figure 3.4a shows that the required brake horsepower can be determined at each flow rate. For liquids other than water, the brake horsepower obtained from the graph must be multiplied by the specific gravity of the liquid.

It should be noted that as flow is increased, the brake horsepower increases due the increased pump loading. As shown in Figure 3.4b, brake horsepower is often represented by contours that represent lines of equal

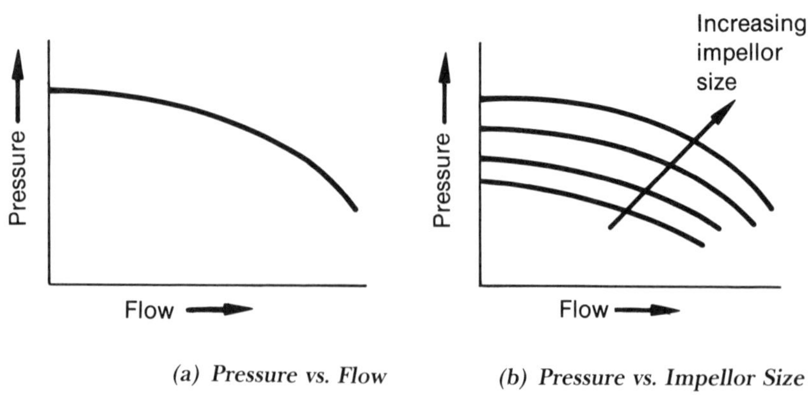

(a) Pressure vs. Flow *(b) Pressure vs. Impellor Size*

Figure 3.3
Pump Performance Curves — Head

brake horsepower requirements. Note that as the operating point moves out on the pump curve, horsepower requirements increase.

Since brake horsepower is a function of impeller size, manufacturers often graph the brake horsepower for each impeller size that can be installed in the pump (see Figure 3.4c).

Efficiency

As the hydraulic energy generated by the pump can be calculated at a given operating point and the energy requirement at the pump shaft (brake horsepower) is also known, pump efficiency, which is the dimensionless

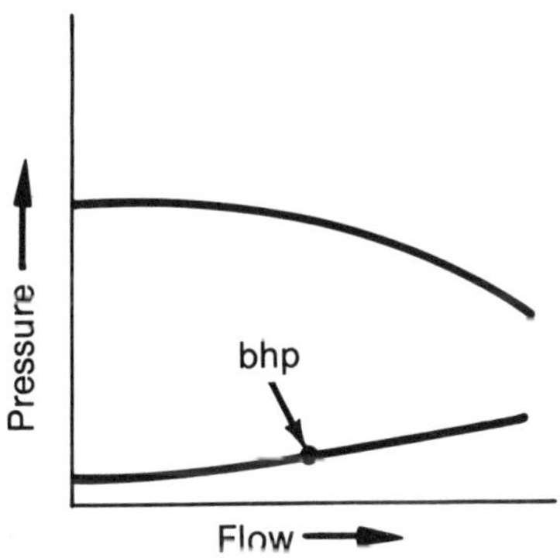

(a) Brake Horsepower vs. Flow

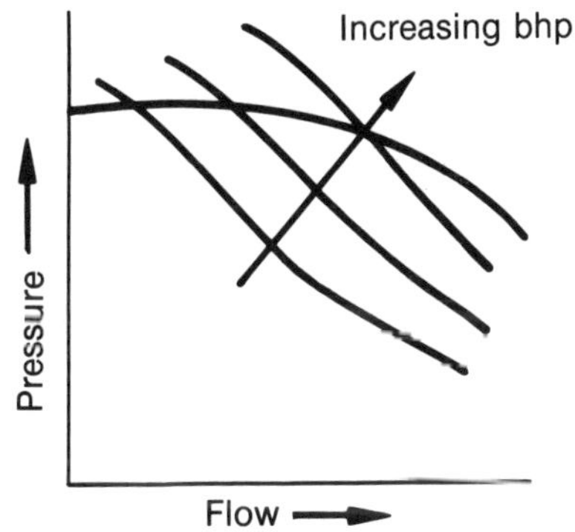

(b) Brake Horsepower vs. Flow

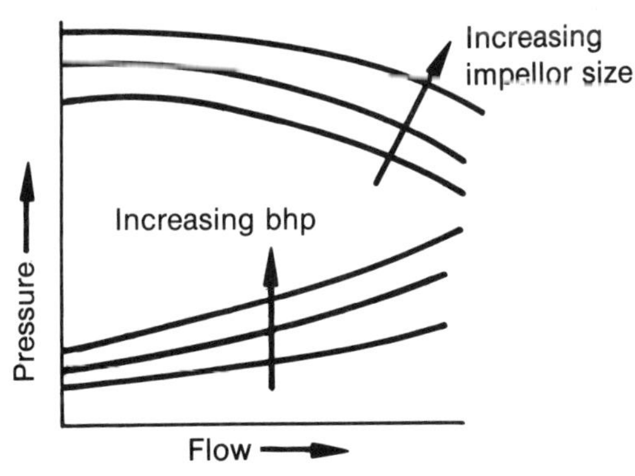

(c) Brake Horsepower vs. Impellor Size

Figure 3.4
Pump Performance Curves — Brake Horsepower

ratio of the hydraulic energy produced to the brake horsepower, can be calculated and represented graphically. As shown in Figure 3.5a, the efficiency can be determined at any flow rate. Notice that the efficiency is zero at no flow conditions, increases as flow increases before peaking, and decreases with increased flow. Pump efficiency data is often expressed as contours representing lines of equal efficiency.

From Figure 3.5b, the pump efficiency at any operating point can be determined. The design operating point of the pump is located preferably in an area of higher efficiency, in this case, at moderately high flows and pressures relative to pump capacity.

Net Positive Suction Head (NPSH)

As the liquid velocity increases through the pump, there may be locations within the pump where the local pressure is below the suction pressure. Should the pressure fall below the vapor pressure of the liquid before rising to the discharge pressure, *cavitation*, the formation and subsequent implosion of bubbles, can occur and cause pump and/or piping damage. Signs of cavitation include pump/piping vibration, a sound as if rocks or ping-pong balls were inside the pump, and an eventual loss of pump output caused by progressive pump damage.

The *required NPSH* ($NPSH_R$) is the amount of suction head expressed in meters (feet) required to keep the liquid from cavitating. NPSH requirements for each impeller are typically presented in graphic form (see Figure 3.6).

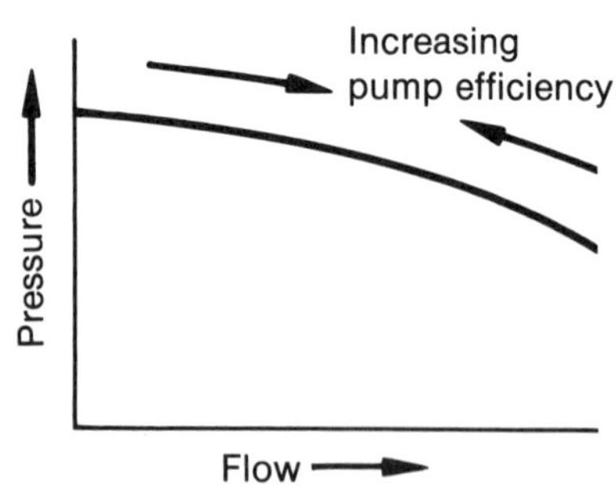

(a) Efficiency vs. Flow

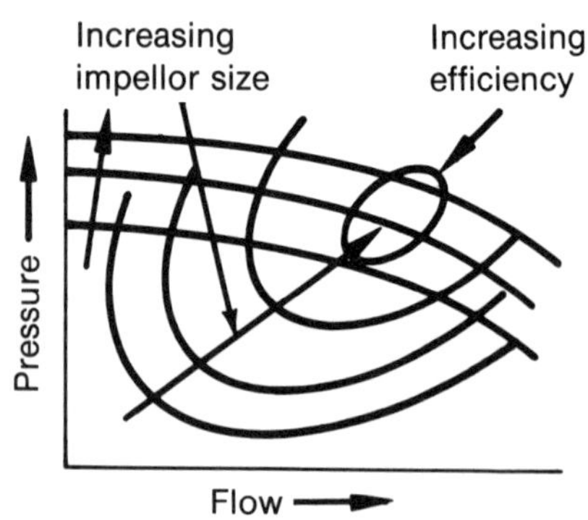

(b) Efficiency Contours

Figure 3.5
Pump Performance Curves — Efficiency

The *available NPSH* ($NPSH_A$) is the sum of the barometric pressure, gage pressure at pump suction corrected to pump centerline, and the velocity head in the suction pipe, less the vapor pressure of the liquid at operating conditions, where all pressures (vacuums) are expressed in meters (feet) of water column.

Composite Performance Curves

To show pump performance data more concisely, the head, brake horsepower, efficiency, and NPSH requirements are typically presented for all impeller sizes on one graph, known as the *composite pump curve* or, more commonly, the *pump curves*. Should the design operating point lie between two of the pump curves, a larger size impeller can be trimmed so that pump performance will coincide with the desired operating point.

In Figure 3.7, note that individual pump curves are presented for 1800 and 3600 rpm motors and that, although the curves are similar in shape, higher flows and pressures are generated by using a higher speed pump. $NPSH_R$ and horsepower requirements are higher at higher operating speeds.

For example, from the pump curves in Figure 3.7, the pump operating at 3600 rpm with a 6½-inch (165 mm) impeller is capable of moving 200 gpm (45.5 m^3/hr or 757 lpm) at a head of approximately 165 feet (50.3 meters). The pump requires approximately 14 bhp (10.4 kW) of input rotational energy and 10 feet (3.0 meters) of $NPSH_R$. The pump operates at an efficiency of approximately 60 percent.

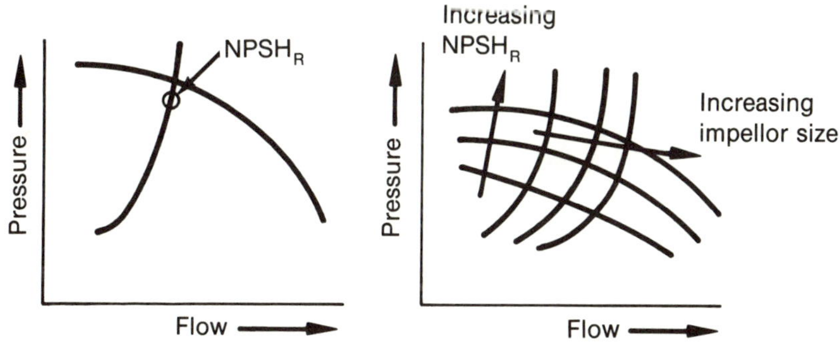

(a) $NPSH_R$ vs. Flow *(b) $NPSH_R$ vs. Impellor Size*

Figure 3.6
Pump Performance Curves — $NPSH_R$

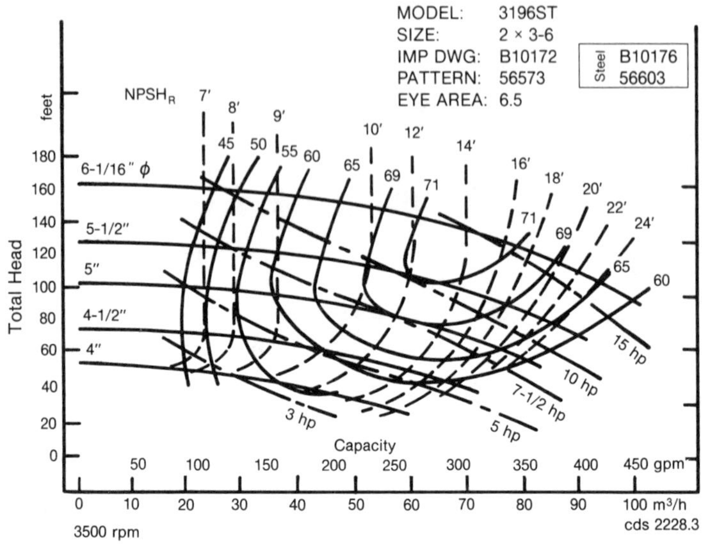

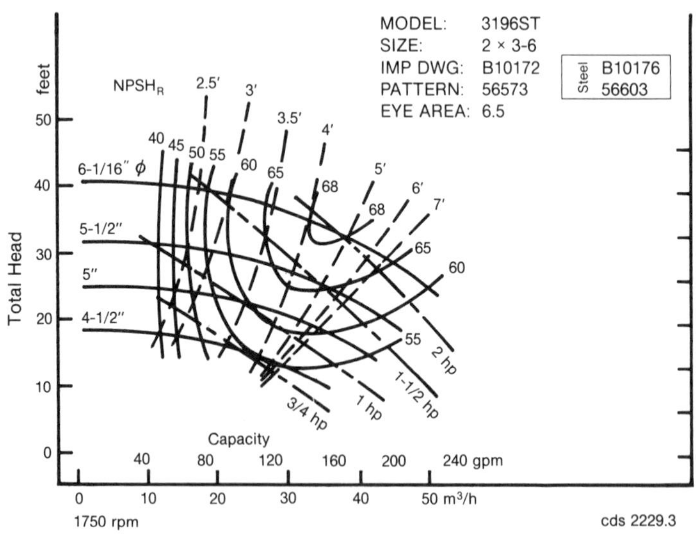

Figure 3.7
Typical Composite Centrifugal Pump Performance Curve

Affinity Laws

The relationship between important pump parameters can be expressed mathematically for centrifugal pumps using the *affinity laws* below where Q is flow, H is head, bhp is brake horsepower and N is pump speed where the pump impeller diameter is held constant.

$$(Q_1/Q_2) = (N_1/N_2)$$

$$(H_1/H_2) = (N_1/N_2)^2$$

$$(\text{bhp}_1/\text{bhp}_2) = (N_1/N_2)^3$$

When the pump speed is held constant, the following relationships are used.

$$(Q_1/Q_2) = (D_1/D_2)$$

$$(H_1/H_2) = (D_1/D_2)^2$$

$$(\text{bhp}_1/\text{bhp}_2) = (D_1/D_2)^3$$

When the pump performance is known at one operating point, the applicable equation can be used to determine pump performance at another operating point. It should be noted that the efficiency remains virtually constant for changes in speed and small impeller diameter changes. For example, if a given pump is operated at 90 percent of its nominal operating speed, its capacity, head, and brake horsepower requirements will be 0.9 or 90 percent, $(0.9)^2$ or 81 percent, and $(0.9)^3$ or 72.9 percent of the original operating conditions, respectively. Note that a 10 percent speed reduction results in a 27.1 percent reduction of brake horsepower.

PARALLEL PUMP OPERATION

As each identical pump will have the same shut-off head, a pump curve for parallel operation is determined by adding the capacities of the pumps at each head. Parallel operation also reduces the $NPSH_R$ requirements on each individual pump. The operating curves for a single pump and two pumps in parallel are shown in Figure 3.8.

SYSTEM PERFORMANCE

The piping system introduces both static and dynamic pressure drops, which will determine where the pump will operate on its curve. A *static pressure drop* is introduced by the elevation changes that the liquid must overcome

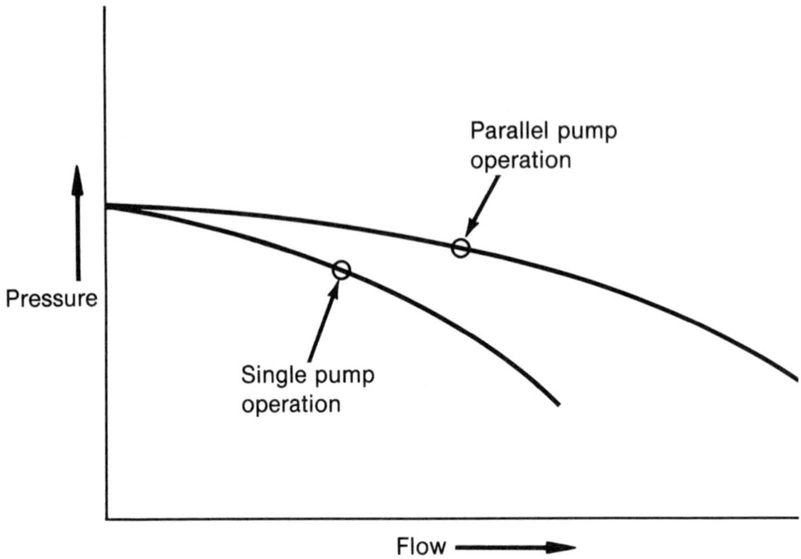

Figure 3.8
Single vs. Parallel Pump Operation

so that flowing conditions can occur. *Dynamic pressure drops* are caused by pipe friction losses, pressure drops across equipment that vary with flow (such as throttled control valves), and the like.

A restriction to flow in the turbulent flow regime (such as pipe friction or a control valve) causes a pressure drop that is proportional to the *square* of the flow through the restriction. Therefore, under these conditions, doubling the flow through the restriction will cause the differential pressure across the restriction to be *four* times as great.

The system head curve can be represented graphically in a manner similar to the pump performance curves (see Figure 3.9a). The upward shift in the curve from the origin at no flow is due to static system head, while the curved shape of the system curve is due to the quadratic relationship between flow and the differential pressure attributable to the piping system.

The intersection of the pump performance curve and the system head curve will determine the operating point of the system and, therefore, where the pump will operate on its curve (see Figure 3.9b).

Static pressure and dynamic pressure drops can affect where the pump operates on its curve, as shown in Figure 3.9c. Also, the operating point of

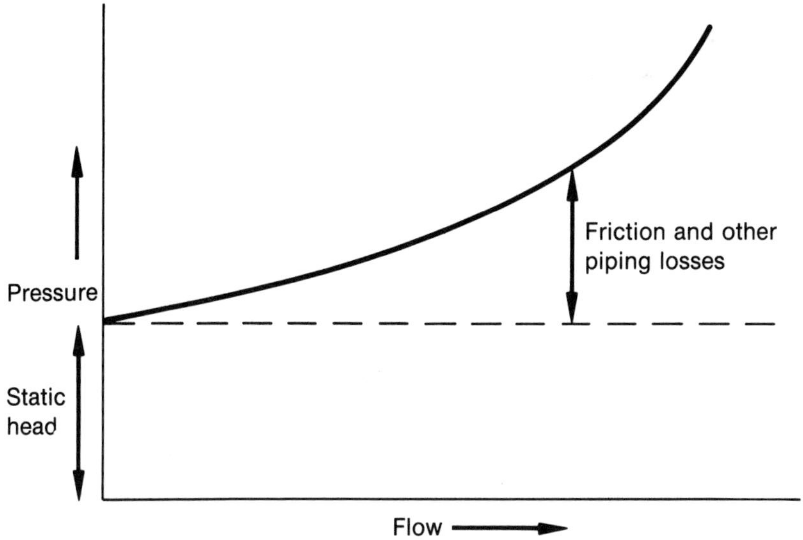

(a) Typical Piping System Head Curve

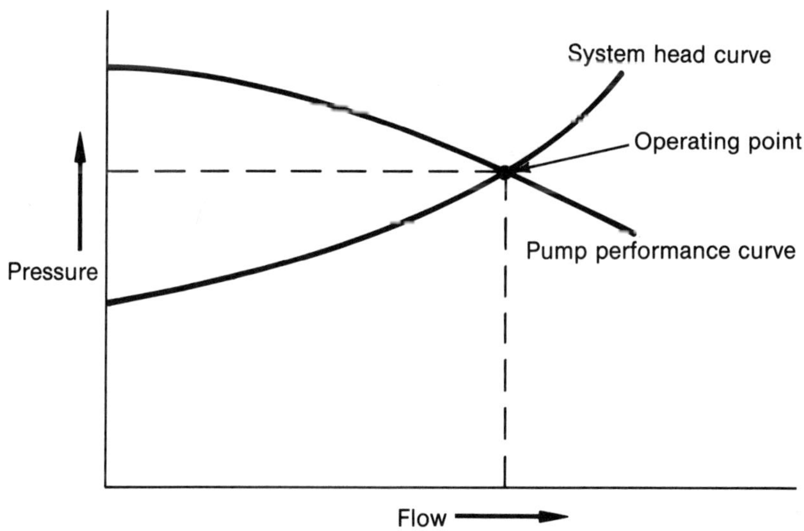

(b) Effect of the Pump Performance Curve on the Operating Point

Figure 3.9
Typical System Head Curves

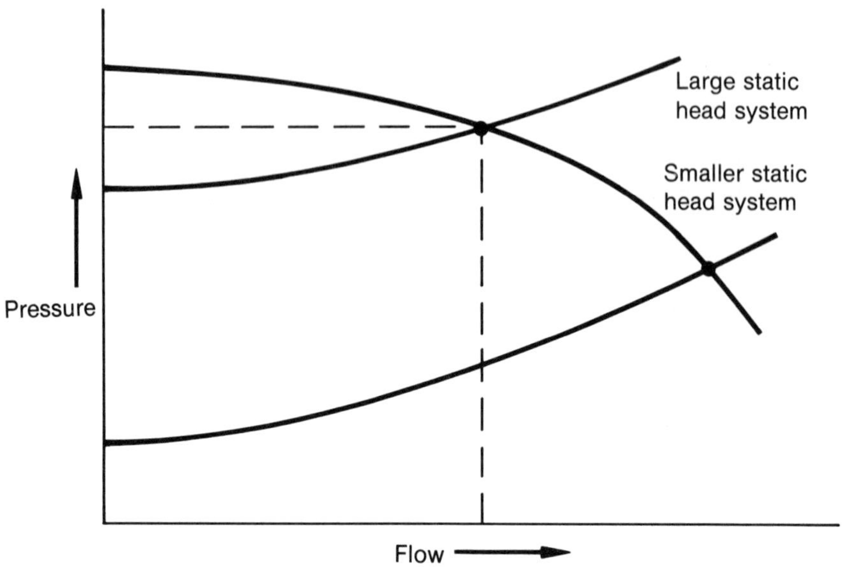

(c) Effect of Static Head on the Pump Operating Point

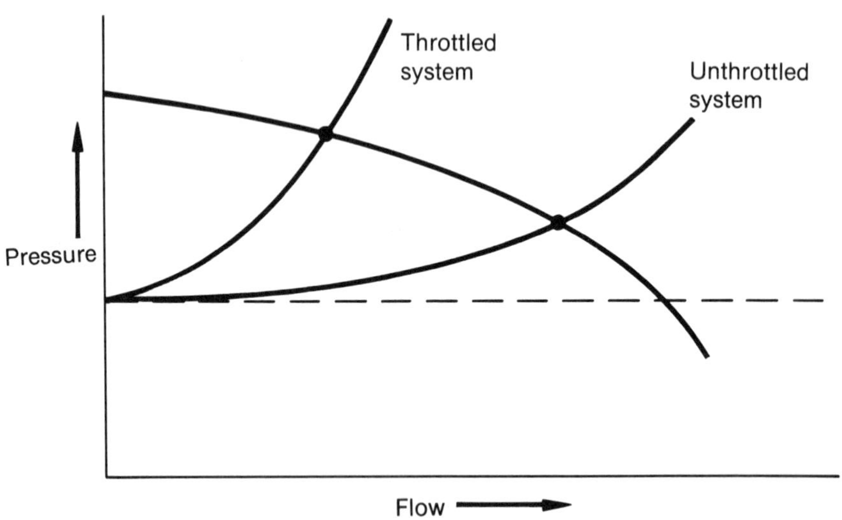

(d) Effect of Dynamic Pressure Drop on the Pump Operating Point

Figure 3.9
(continued)

the system will slide up and down the pump curve in response to changes in the dynamic pressure drop (see Figure 3.9d).

CENTRIFUGAL FANS AND BLOWERS

The operating characteristics of centrifugal fans and blowers are essentially identical to those of centrifugal pumps with the following notable exceptions:

(1) In many applications, the static system pressure is zero.
(2) Centrifugal fan and blower curves can be relatively flat.

CONSTANT TORQUE LOADS

Whereas the brake horsepower of centrifugal equipment increases as the cube of the load (impeller size), *constant torque* loads exhibit an approximately linear relationship between brake horsepower and load. Many pieces of non-centrifugal equipment exhibit what can be approximated as constant torque behavior. Examples of constant torque loads include positive displacement pumps, conveyor belts, positive displacement compressors, and the like.

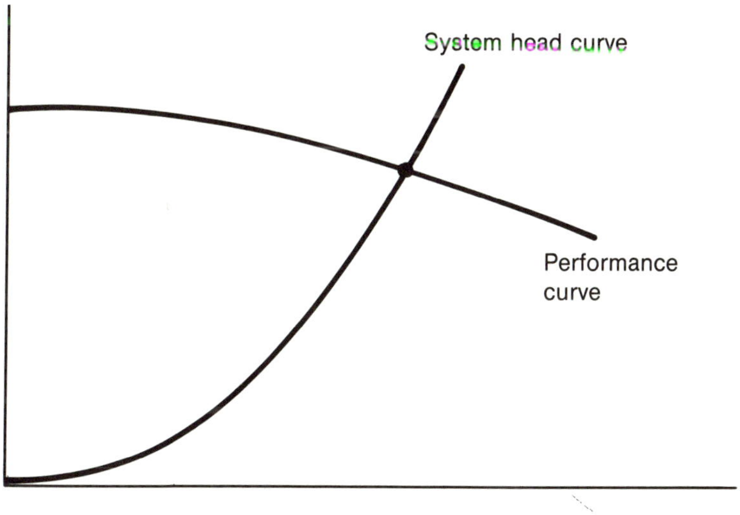

Figure 3.10
Typical Centrifugal Fan and Blower Curve

Chemistry Basics

4

Quantification of the effects of process modifications often entails the understanding and application of the principles of chemistry and/or the related disciplines of chemical engineering, combustion and metallurgy. The ability to adequately describe modifications in a scientific and logical manner often determines whether or not certain changes are operationally, economically, or environmentally advantageous.

It must be understood that plant modifications, however small they may appear, can and do drastically affect other facets of plant operation. As a prerequisite to a successful project, all of the effects of a particular modification must be addressed. Changes that are described using the above-mentioned principles often result in operational, economic, and environmental benefits, which can form a quantifiable basis for modification of the process.

Some of the principles of chemistry, chemical engineering, combustion and metallurgy are described here in a cursory manner to impart a basic understanding of some practical aspects of the subjects.

CHEMISTRY

Chemistry is the study of substances, their composition and how they interact with each other. A *substance* is composed of molecules, which are the smallest part of a substance having all of the properties of the substance. A *molecule* is an interconnection of atoms of individual elements. Each molecule has a *molecular weight* equal to the sum of the molecular weights of each atom comprising the molecule. The molecular weight of some elements are listed below.

Element	Symbol	Molecular Weight
Hydrogen	H	1
Carbon	C	12
Nitrogen	N	14
Oxygen	O	16

Hydrogen, nitrogen, and oxygen are diatomic and exist as paired atoms when not chemically bound to form other molecules. In general, a molecule consists of identical combinations of atoms of different elements, which can be described by a chemical formula. For example, water consists of two hydrogen atoms and one oxygen atom. The chemical formula for water is H_2O and its molecular weight is $(2 \times 1) + 16$, or 18. Similarly, carbon dioxide (CO_2) has a molecular weight of $(2 \times 16) + 12$, or 44.

The Law of Conservation of Matter states that mass is neither created nor destroyed in ordinary chemical changes. Therefore, when two substances react, the mass of the reactants is equal to the mass of the products of the reaction. While the atoms of different molecules recombine with each other as a result of a chemical reaction, the atoms themselves remain intact during the reaction. Therefore, the mass of each respective element in the reactants is equal to the mass of that element in the products of the reaction. Using these principles, reactions can be described using chemical equations. For example, the equation for the burning of methane (the primary constituent of natural gas) and air to form carbon dioxide and water is given by the following equation:

$$CH_4 + 2\ O_2 \rightarrow CO_2 + 2\ H_2O$$

Note that the number of molecules of each element is the same on each side of the equation, that oxygen is diatomic in its pure form, and that some elements in the air (predominantly nitrogen) do not react and are not shown in the equation.

One *gram-mole* (g-mol) of a substance has a mass (in grams) numerically equal to the molecular weight of the molecule. Regardless of the substance, each gram-mole contains the same number of molecules and occupies the same volume at standard conditions of pressure and temperature (0 °C and 760 mm of mercury absolute). Writing the molecular weights below each reactant and reaction product in the above equation results in the following equation:

$$CH_4 + 2\ O_2 \rightarrow CO_2 + 2\ H_2O$$

16 64 44 36

Note that the sums of the molecular weights on each side of the equation are equal. Interpreting the equation, 16 grams of methane and 64 grams of oxygen react to form 44 grams of carbon dioxide and 36 grams of water. The ratio of the mass of each of the molecules will be in the proportion given by the molecular weights. For example, 64/16, or 4 grams of oxygen are required for each gram of methane. Similarly, 36/16, or 2.25 grams of carbon dioxide are formed for each gram of methane consumed. Note that nitrogen and other inert gases present in the combustion air are heated as a result of this reaction.

SPECIFIC HEAT

The *specific heat* of a substance (c_p) is the amount of heat necessary to raise the temperature of a unit mass of a fluid by one degree. The specific heat of a substance will vary as a function of temperature and pressure and is usually considered a constant unless large variations in temperature and/or pressure occur. This concept is described by the following relationship used to calculate the amount of heat gained (lost) by a given mass of fluid due to a change in temperature:

$$Q = m \times c_p \times \Delta T$$

where Q is the amount of heat, m is mass, c_p is the specific heat and T is the change in fluid temperature.

For example, the heat input required to raise the temperature of 1000 lb/hr of a gas with a specific heat of 0.50 Btu/lb-°F by 300 °F is (1000 lb/hr × 0.50 Btu/lb-°F × 300 °F), or 150,000 Btu/hr. Similarly, the amount of heat required to raise the temperature of 1000 kg/hr of a gas with a specific heat of 0.25 cal/g-°C by 200 °C is (1.162 W/cal × 1000 kg × 0.25 cal/g-°C × 200 °C), or 58100 W. It should be noted that the specific heat expressed in the English system as Btu/lb-°F is numerically equivalent to its metric counterpart of cal/g-°C.

HEAT OF VAPORIZATION

The *heat of vaporization* of a liquid is the amount of energy per unit mass required to effect the change of state from liquid to gas.

For example, the heat of vaporization of water at atmospheric pressure is 970.3 British thermal units (Btu) per pound (0.6269 kWh/kg). In other words, 970.3 Btu (0.6269 kWh) of thermal energy is required to vaporize each pound (kilogram) of water at 212°F (100°C) and 14.696 psia (1.01325 bar).

COMBUSTION

In practical terms, *combustion* can be defined as the oxidation of a fuel that results in a release of energy in the form of heat. The amount of heat released during the reaction, that is, the higher heating value (HHV) of the fuel, is used to vaporize and heat the products of combustion. Often, the heating value of a fuel is expressed in terms of its lower heating value (LHV), which is the higher heating value of the fuel less the heat necessary to vaporize the products of combustion. The LHV is the amount of energy that is available to elevate the temperature of the products of combustion.

As an example, in the equation above describing the combustion of methane, 36 kg of water are produced as a result of the combustion of 16 kg of methane. Therefore, 36/16 or 2.25 kg of water vapor are produced per kg of methane reacted. The LHV, calculated by subtracting the heat of vaporization of the water from the HHV of 1 kg of methane (14.3 kWh/kg), is

$$\text{LHV} = 1\ \text{kg} \times 14.3\ \text{kWh/kg} - 2.25\ \text{kg} \times 0.6269\ \text{kWh/kg} = 12.9\ \text{kWh}$$

The amount of energy necessary to vaporize water is a function of the hydrogen content of the fuel. In the case of methane, approximately 10 percent of the HHV of the fuel is used to vaporize water produced by the reaction, such that the LHV of natural gas (assumed to be methane) having an HHV of 1000 Btu/ft^3 and a specific gravity of 0.6, can be calculated as:

$$\text{LHV} = \left\{ (1\ \text{lb}) \left[\frac{\dfrac{1000\ \text{Btu}}{\text{ft}^3}}{(0.6)\ (0.075\ \text{lb/ft}^3)} \right] - (2.25\ \text{lb}) \left(970.3\ \frac{\text{Btu}}{\text{lb}} \right) \right\}$$

$$= 902\ \frac{\text{Btu}}{\text{ft}^3}$$

COMBUSTION EFFICIENCY

Combustion efficiency can be defined as the ratio of the useful energy removed from products of combustion to the HHV of the fuel. In a typical steam boiler, the combustion efficiency is the ratio of the net heat gain of the water/steam to the HHV of the fuel input.

Stated differently, combustion efficiency is the ratio of useful energy output to the amount of energy input. In boiler applications, the most common combustion processes, combustion efficiency is typically determined by using a combustion efficiency table for the applicable fuel, which relates combustion efficiency to the excess oxygen in the stack gas and the difference between stack temperature and ambient temperature. Tables are commonly used because most boilers do not have sufficient measurement devices to use the equations presented above, since such measurements are not required for operation.

Where tables do not exist for the desired fuel or operating conditions, the combustion efficiency can be calculated as

$$\frac{\text{HHV} - \text{Heat Losses}}{\text{HHV}}$$

where the heat losses are calculated using the relationships described above. Note that stack losses, radiation losses, losses due to leakage, and the like should be accounted for in the above equation, although the use of a combustion efficiency table accounts for only stack losses.

Interpreting the above analysis, when it is desired to heat a fluid using the combustion process, the fuel must not only heat the substance, but also vaporize and heat the products of combustion for eventual discharge, subsequently resulting in a heat loss. The fuel input will be greater than the amount of heat absorbed by the fluid. Given the HHV of the fuel, the amount of fuel needed to generate the required heat can be calculated as

$$\text{heat input} = \frac{\text{heat output}}{\text{combustion efficiency}}$$

METALLURGY

An allowance is typically incorporated into the design of metallic components so that the equipment will have a desired useful life when operated under the design operating conditions. Phenomena typically discovered during operation, such as excessive corrosion, flashing, cavitation, abrasion, thermal cycling, and thermal shock, can drastically reduce the useful life of metal components. A reduction of either the amount of mechanical stress or the abruptness of change of mechanical stress will increase the anticipated equipment life and reduce the number of equipment failures, although the amount of this reduction is not generally predictable.

Instrumentation Basics

5

Instruments are applied to industrial processes to remotely sense and display various process parameters. The information provided forms a part of the basis from which decisions regarding the operation of the process are made. Such decisions may be made manually after examining the data or by utilizing measurement and control devices automatically. The ultimate goal of applying instrumentation and associated controls to a process is to achieve safe, stable, reliable, and economical plant operation.

PROCESS CONTROL

A generalized process with associated measurement and control devices can be modeled as shown in Figure 5.1. Such a model is composed of the process with associated disturbances, a measurement of a parameter that is indicative of process operation, a control device that compares the desired value of the parameter with the measurement, and the final control device, which modifies the operation of the process. When many measurements and final control elements are involved, the control can utilize complex strategies.

PROCESS

A *process* is anything that changes. Processes can be misleadingly simple, such as driving a car, or complex, such as the production of a specialized

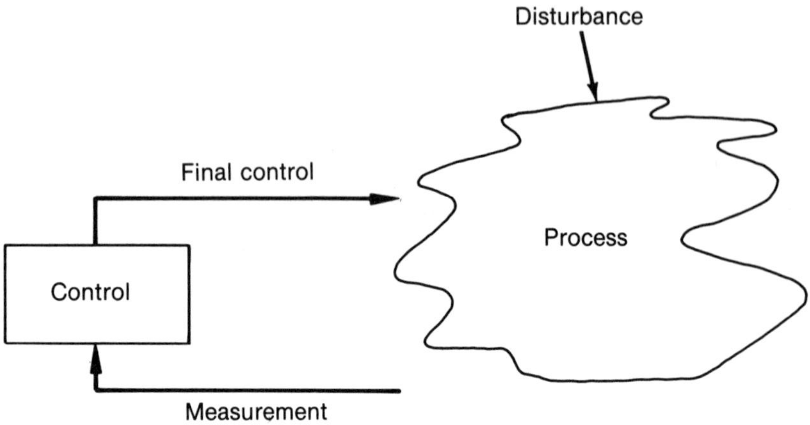

Figure 5.1
Generalized Process Model

chemical. Regardless of the complexity of a modification or an addition to a process, it is imperative that the affected portion of the process be completely understood in order to determine the technical and economical feasibility of the change. Before such a modification is made, ramifications of the change should be examined with extreme care, preferably using formal tools such as fault tree or "what if" analysis, to determine whether the change will result in any undesirable or unsafe conditions under certain operating conditions.

MEASUREMENT

The selection of an instrument to measure the proper parameter accurately and reliably is extremely important and is often the key to the success of a process modification. As many devices are available to measure process parameters, care should be taken to select an instrument that is compatible with the process and appropriate for the application.

The most common parameters used in industrial applications are flow, level, pressure, and temperature, although there are many other parameters that may be measured.

Flow Measurement

Flow can be defined as the amount of a fluid that passes per unit time. In general, flow measurement devices either infer flow or measure volume, velocity, or mass. When considering flow measurement, it is usually important to determine which type of flow measurement is desired and to select a flowmeter accordingly. Care should be taken to avoid the many pitfalls encountered during the flowmeter selection procedure.

For example, a mass flow measurement might be preferred to measure the feeds to a reactor. When excessive fluid velocity will overload equipment, a flowmeter that measures velocity may be applicable to operate the equipment as close as possible to its capacity, without overloading the equipment.

Level Measurement

Various level measurement devices are available that either measure or infer level. When level measurement is considered, it is often important to determine which type of level measurement is desired and to select a level transmitter accordingly. Level measurements are often hampered by turbulence, foaming, interfaces, and other difficulties.

For example, a level transmitter that measures true liquid level may be used to determine the level in a tank. However, the same transmitter would not be appropriate when the object of the level measurement is to determine the height of foam in the tank or perhaps the interface between two liquids in the tank.

Pressure Measurement

Pressure transmitters are available to measure virtually any range of pressure and/or vacuum. Care should be taken when selecting a pressure transmitter due the corrosion considerations of the relatively thin diaphragms required for the most common ranges.

Temperature Measurement

Various application-specific techniques are available to measure temperature; however, it should be noted that thermal and other effects present in real applications often limit the accuracy of measurement.

For example, the temperature of a fluid in a pipe may be measured by inserting a temperature sensor in a well into the pipe, while this technique would not be applicable to accurately measure the surface temperature of semi-molten glass.

Other Process Measurements

Transmitters are available to measure many process parameters such as dew point, humidity, pH, conductivity, vibration, speed, chemical analysis, and position, to name but a few. Care should be taken when selecting these and other instruments to assure that the instruments are compatible with the process at hand.

CONTROL DEVICES

Typical industrial *control devices*, called controllers, are those that compare the desired value of a parameter with its measurement and generate an appropriate output signal. Loops with controllers are termed PID loops when the controller output is a function of the error (proportional), the amount of time the error exists (integral), and the rate of change of the error (derivative). While PID controllers are most often utilized for process control, other control strategies do exist. Controllers can be implemented individually or as part of a larger computer-based system.

FINAL CONTROL ELEMENTS

Final control elements are those devices used to manipulate the process in accordance with the output signal generated by the control device. The final control element most commonly used in process industries is the control valve, although other final control elements such as those that regulate position or speed are also available.

The compatibility, reliability, and response of the final control element as it relates to the process and other control devices are important factors in a successful installation. The energy efficiency of the final control element as it relates to the process can be an important factor in the economic operation of the process.

CONTROL LOOP

The *control loop* consists of all of the equipment described above and its interconnection to the process. The function of a control loop is to measure a process parameter and control that parameter at the desired value. Achieving this goal is a function of how well suited each part of the control loop is to the process.

Addition of a new control loop or modification of an existing control loop should be handled carefully to avoid unnecessary pitfalls. Selection of the process parameter to be controlled should be based upon detailed process knowledge, while the selection of measurement instrumentation, control devices, and final control elements should be performed by instrumentation specialists when possible. Failure to control the proper parameter or to select the proper equipment can doom a project to complete failure.

SAFETY

The safety of the control loop is paramount to its successful implementation. Formal safety analysis techniques, such as fault tree and "what if" analysis, can be used to investigate the possibility of failure modes that may be unsafe. These techniques may also uncover undesirable operating characteristics of a proposed control loop.

Prime Utility Costs

6

The perception that utilities are available gratis still prevails in some commercial and industrial facilities, and there appears to be no end in sight for this myth. It is the rule, rather than the exception, that utility services are continually being wasted during the day-to-day operation of commercial and industrial facilities. Operation of plants under the premise that "the plant operates where it is most convenient" has propagated itself throughout industry.

Fighting this syndrome requires a minimum of two areas of emphasis. The first step is to make all parties, especially utility users, aware of the real cost of utilities in an attempt to develop a sense of appreciation of the magnitude of the costs involved. Often, it then becomes easier to decide when it might be advantageous for the plant *not* to operate where it is most convenient, but rather where it is most economical. The second area is a longer term commitment to modify existing plants and to design new facilities under the premise that, since the plant will operate where it is most convenient, the most convenient operation should occur (by design) where the operation is most economical.

Calculation of utility costs may appear to be a rather simple matter at first; however, seemingly trivial calculations, such as interpreting an electric bill, can become amazingly complex to the casual observer.

ELECTRICITY

The electric user should be intimately attuned to the cost of electricity and perhaps, more importantly, to what the user is actually paying for. Each utility has many different rate structures to describe the rules that will be used to calculate the cost of electricity. While the residential customer has no choice of rate structure, the industrial user may have flexibility in the choice and hence be able to select the rate structure that provides adequate electric service at the minimum cost.

Electrical Billing Terminology

Terms commonly used to describe rate structures and billing are explained below to develop a conceptual understanding of each principle involved. Differences in terminology between utilities in different locales may be noted; however, the concepts presented are almost universally applicable.

Time-of-Day Billing

The typical electric utility experiences its peak electrical load during certain hours on certain days during the different seasons. The utility must maintain sufficient generating capacity to handle these peaks even though this capacity may sit idle much of the year. To recover the capital and increased operating costs associated with this less efficient equipment, as well as to encourage users not to use electricity during these peak periods, commercial and industrial users are usually subject to time-of-day billing.

Typically, rates during the fall, winter, and spring are higher during weekday working hours than for the remaining hours of the week. The intent is to encourage large users to consume electricity in off-peak hours to smooth out the load on the utility throughout the 24-hour day or, conversely, to charge a premium to those users who use electricity during peak hours. Summer rates are typically divided into weekday daytime, weekday evening, and night/weekend rates, each successive period having lower rates.

Energy Charges

Of great importance to the electric user is the means by which electric usage is measured and the billing for that usage, often termed the energy charges. Electrical consumption is measured with an electric meter that measures power consumption in kilowatt hours (kWh). When time-of-day billing applies to the user, the electric meter totalizes energy consumption individually for each time period.

Demand Charges

In addition to the energy charge, the utility imposes a demand charge in an attempt to equitably recuperate the capital costs associated with distributing electricity to each user. The demand charge is typically based upon the highest power consumption made during a 15-minute period during the billing period. Demand charges are usually different for each time-of-day billing period and are billed for each period in effect.

Power Factor Penalty

Some rate structures state that the demand charge can be no less than a certain percentage (typically 90 percent) of kVA demand, as opposed to kWh demand. As most users have power factors of less than the stipulated percentage, this effectively makes the user pay demand charges based upon power factor. Improvements in plant power factor will reduce these charges.

Hatchet and Ratchet Clauses

Hatchet/ratchet clauses appear in some rate structures and are a method by which demand charges for a given billing period are determined. Typically, demand charges will be the higher of the actual demand during the billing period or a percentage of the highest demand period during either the previous year or the previous summer.

The hatchet refers to the abrupt rise in the demand charges when demand is higher than on recent bills, while the ratchet refers to the reduction in demand charges when demand is lower than on recent bills.

Interruptible Service

Electric service is typically uninterruptible (firm); however, interruptible service may be available, usually at lower cost. Interruptible users must disconnect their loads from the power grid within a certain period of time after notification by the utility. This helps the utility minimize the production and/or purchase of high priced electricity by shedding large users during periods of peak electrical consumption. The interruptible customer typically has alternate generating capacity that can be used during these periods.

Incremental Electric Costs

Of apparent interest to the electric user is the cost of electricity obtained by dividing the cost of the power by the amount of power consumed. However, of much greater importance is the incremental cost of electricity. The poignant question is not what electricity costs per kWh, but rather, the cost of the last, or incremental, kW used or saved, which allows definitive quantification of the effects of additions and deletions of plant load on the electric bill. These costs should be duly calculated by the electrical user at each plant location.

Examples of Electric Rate Structures

As the sampling of typical rate structures presented below can and do vary with location, the following analysis should be done for each plant. There are often miscellaneous corrections to the bill, such as fuel surcharges, taxes, and the like, which are approved by the local Public Utility Commission (PUC). These charges or credits cannot be predicted and are hence not included in these calculations.

Residential

Residential electric rate structures apply to homeowners and apartment dwellers. These rate structures typically consist of a fixed cost per kWh, subject to a minimum charge for the use of the electric meter.

Consider a rate structure in which the monthly bill reflects a customer charge of US$5.00 and energy charges of US$0.10/kWh on the first 500 kWh and US$0.09/kWh for all kWh in excess of 500 kWh. Assuming that over 500 kWh are consumed each month, the annual cost of an incremental kW of electricity is US$0.09/kWh × 8760 hours per year, or US$788.40 per kW-year. The significance of this figure is that when an additional 1 kW of load is added (deleted) for the entire year, the electric bill will increase (decrease) by US$788.40 per year, or US$0.09 per kWh.

Residential bills usually feature the highest cost per kWh due to the relatively large capital investment and maintenance required to reliably distribute small quantities of power to a large number of users.

Industrial

Industrial electric rate structures typically apply to small to medium size electric users who require 120- and 240-volt service.

Consider the outline of the industrial rate structure and note the complexity of the calculation of the incremental cost of electricity, assuming 15,000 kWh billing demand per month and that winter demand is less than 70 percent of summer demand.

Typical Industrial Rate Structure

Rate — Monthly	Summer*	Other
(1) Customer Charge	$5.00	$5.00
(2) Demand Charge		
First 5 kW or less	No Charge	No Charge
All over 5 kW	$10.00/kW	$6.00/kW
(3) Energy Charge		
First 1,250 kWh	$0.10/kWh	$0.08/kWh
Use up to 30,000 kWh or 300 hours use of billing demand, whichever is greater	$0.06/kWh	$0.06/kWh
Use in excess of 30,000 kWh or 300 hours of billing demand, whichever is greater	$0.045/kWh	$0.045/kWh

Special Provisions

Separately Metered Space Heating

All kWh @ $0.06/kWh (October through May)

All kWh @ $0.10/kWh (June through September)

(4) Plus: Fuel Adjustment Charge

(5) Plus: Increase in Rates and Charges

Customer Transformer Ownership

When a customer provides all equipment required to take service at a primary voltage, the demand charge stated above shall be reduced by $1.00 per kW of demand billed.

June → September

Energy:	(30 + 31 + 31 + 30 days) (24 hrs/day) ($0.06/kWh)	$175.68
Demand:	(4 months) ($10.00/month)	40.00

October → May

Energy:	(31 + 30 + 31 + 31 + 28 + 31 + 30 + 31 days) (24 hrs/day) ($0.06/kWh)	349.92
Demand:	(8 months) ($6.00/month) (70%)	33.60

Subtotal = $599.20

$$\textit{TOTAL} = \frac{599.20}{0.95\ \text{(local tax)}} = \$630.73$$

**Summer Billing Months:* June through September for bills rendered monthly.

Determination of Demand

For demands in excess of 5 kW the monthly billing demand in kW shall be either the greatest connected load or the greatest 15-minute integrated demand, determined as follows:

Billing demand may be on a connected load basis when a demand meter would not reduce the billing demand, the installation is temporary, or the device has a large instantaneous or highly fluctuating demand.

Billing shall be on a demand meter basis in all other cases and shall be billed at not less than 90 percent of the kVA demand. The billing demand for the billing months of October through May inclusive shall not be less than 70 percent of the highest metered demand for the preceding billing months of June through September inclusive.

Customers who are not expected to use at least 5 kW will not normally have a demand meter installed. However, should such customer use 800 kWh or more in each of two consecutive months, the Company will install a suitable demand meter.

The incremental cost of 1 kW of electricity is US$630.73 per year or US$0.072 per kWh. Note that demand charges can be reduced if the user can take service at primary voltage, and that winter demand charges are based upon summer demand, so that a reduction in the peak summer demand can result in electrical cost savings throughout the year. This is one form of a hatchet/ratchet clause.

Industrial Time-of-Day

Primary industrial rate structures are available to electric users who meet minimum usage requirements and accept primary power (typically 13.2 kV or higher) from the utility. This rate structure usually features time-of-day provisions but typically results in lower kWh and incremental kW costs for uninterruptible firm power.

Typical Time-of-Day Primary Industrial Rate Structure

Rate — Monthly

Any general primary service customer who maintains a minimum demand of 1,000 kW during any two of the previous twelve months may be eligible for this service classification. A customer who fails to maintain a demand of at least 1,000 kW during any two of the previous twelve months may, at the customer's option, transfer to another service classification provided that such transfer shall only be made on an annual anniversary date that such customer began service hereunder.

(1) Demand Charge

Period I	— All kW	$6.00/kW
II	— All kW	$4.00/kW
III	— All kW	$3.00/kW
IV	— All kW	No Charge

(2) Energy Charge

Period I	— All kWh	$0.060/kWh
II	— All kWh	$0.054/kWh
III	— All kWh	$0.060/kWh
IV	— All kWh	$0.045/kWh

(3) Plus: Fuel Adjustment Charges

(4) Plus: Increase in Rates and Charges

Definition of Rating Periods

Period I — 1:00 p.m. to 7:00 p.m. prevailing time. Monday through Friday except holidays, June 1 through September 30.

II — 8:00 a.m. to 1:00 p.m. and 7:00 p.m. to 11:00 p.m. prevailing time. Monday through Friday except holidays, June 1 through September 30.

III — 8:00 a.m. to 11:00 p.m. prevailing time. Monday through Friday except holidays, October 1 through May 31.

IV — 11:00 p.m. to 8:00 a.m. prevailing time. Monday through Friday, all hours on Saturdays, Sundays and holidays, all months.

For purposes of this section holidays are: New Year's Day, Memorial Day, Independence Day, Labor Day, Thanksgiving Day and Christmas Day.

Determination of Demand

The billing demand shall be defined as the highest 15-minute integrated kW demand determined during each rating period by the use of a suitable demand indicator, but not less than 90 percent of the kVA demand.

		hrs/wk	*$/kWh*	*$/kW-wk*
June → September				
Energy:				
Period	I	(6×5) = 30	0.060	1.80
	II	(9×5) = 45	0.054	2.43
	IV	(9×5)+(2×24) = 93	0.045	4.19
				8.42
				×17.43 wks
				146.76
Demand:				
Period	I		6.00	
	II		4.00	
	IV		—	
			10.00 × 4 mo.	40.00
Subtotal, June → September				186.76
October → May				
Energy:				
Period	III	(15×15) = 75	0.060	4.50
	IV	(9×5)+(2×24) = 93	0.045	4.19
				8.69
				×34.71 wks
				301.63
Demand:				
Period	III		3.00	
	IV		—	
			3.00 × 8 mo.	24.00
Subtotal, October → May				325.63
June → September				186.76
Subtotal				512.39/kW-yr

$$\text{Total} = \frac{512.39}{0.95 \text{ (local tax)}} = \mathbf{\$539.36/kW\text{-}yr}$$

The incremental cost of 1 kW of electricity is US$539.36 per year or US$0.0616 per kWh. Note that an improvement in power factor can reduce the amount of the demand penalty, and that this particular rate structure bases demand charges on kW demand during the billing period, and that it contains no hatchet/ratchet clauses.

Take-or-Pay Contracts

Contracts are often drawn up between the user and the utility whereby the user agrees to purchase an amount of electricity from the utility. In a take-or-pay contract, this amount of electricity must be paid for whether or not the electricity is actually used.

This type of contract is often used where the utility desires a guaranteed income to cover the cost of capital expenditures necessary to provide electric service to the user. This type of contract may apply where a plant expands beyond the capacity of the utility, requiring a utility expansion as well.

Miscellaneous Rate Structures

Many other rate structures and variations of the above rate structures exist to define billing for each type of user within the utility service area. Examples include:

Interruptible
Three-phase Residential Service
Residential Heating
Residential Time-of-Use
Area Lighting
Purchase of Electricity by the Utility
High Tension Service

NATURAL GAS

Most industrial plants depend on natural gas for a large portion of plant energy requirements. Natural gas is easily distributed and environmentally clean when properly burned.

Natural Gas Billing Terminology

The terminology associated with natural gas rate structures is virtually the same as that for electric utility rate structures.

Examples of Natural Gas Rate Structures

As the sampling of typical rate structures presented below can and do vary with location, the following analysis should be done for each plant. These rate structures are based on uninterruptible (firm) service and will be used for subsequent calculations. There are often miscellaneous corrections to the bill, such as fuel surcharges, taxes, and the like, which are approved by the local Public Utility Commission (PUC). These charges or credits cannot be predicted and are hence not included in the calculations.

Residential

Residential gas rate structures apply to homeowners and apartment dwellers. These rate structures typically consist of a fixed cost per unit volume, typically hundreds of cubic feet (ccf), thousands of cubic feet (mcf), or cubic meters (m^3), subject to a minimum charge for the use of the gas meter.

Consider a rate structure in which the monthly bill reflects a customer charge of US$5.00 for the first 2 ccf (5.66 m^3), US$0.60 per ccf (2.83 m^3) for the next 48 ccf (278 m^3), and US$0.50 for each additional ccf. Assuming that approximately 20 ccf are consumed each month, the incremental cost of natural gas is US$0.60 per ccf. The significance of this figure is that for each ccf increase (decrease) in consumption, the gas bill will increase (decrease) by US$0.60.

Commercial/Industrial

Commercial/industrial rates typically apply to most users, with the exception of individual residences. This rate structure usually consists of a tiered pricing system similar to the residential rate structure above; however, there may be added complexities.

Consider a rate structure in which the monthly bill reflects a customer charge of US$5.00 for the first 2 ccf, US$0.60 per ccf for the next 98 ccf, US$0.45 per ccf for the next 5000 ccf, and US$0.40 for each additional ccf. Assuming that approximately 1000 ccf are consumed each month, the incremental cost of natural gas is US$0.45 per ccf.

Miscellaneous Rate Structures

Other rate structures and variations of the above rate structures exist to define billing for each type of user within the utility service area. Examples include:

Interruptible
Cogeneration Interruptible
Interruptible Priority
Off-Peak
Building and Heating

FUEL OIL

Different qualities of fuel oil can be purchased, ranging from highly refined light oils to high-pour residual oils, each with various amounts of impurities and different heating values. Fuel oil is typically delivered by truck, railroad, barge, or pipe to a plant. The cost of fuel oil should reflect the price of the fuel in addition to transportation costs, local taxes (if applicable), and other miscellaneous charges.

For example, the cost of fuel oil, subject to a 2 percent local tax, which costs US $0.50 per gallon including delivery, is 1.02 × US $0.50, or US $0.51 per gallon.

COAL

Different qualities of coal can be purchased, ranging from low sulfur coal to run-of-mine coal with rocks and other impurities. Coal is typically delivered by railroad or barge but may be transported by truck or piped to a plant. The cost of coal should reflect the price of the fuel in addition to transportation costs, local taxes (if applicable), and other miscellaneous charges.

WASTE FUELS

Since many plants generate substantial quantities of wastes from which it is impractical or uneconomical to produce a final product, waste fuels and gaseous byproducts can compose a substantial portion of an industrial plant's fuel consumption. A plant with suitable combustion and environmental equipment will typically burn wastes for their energy content, thereby displacing purchased fuel and avoiding disposal costs.

The economic value of the waste is the cost of the fuel the waste fuel displaces and is usually lower than the actual value of the raw materials and/or final products that comprise the waste.

Depending upon location, it may be possible to purchase wastes from nearby plants to burn as fuel in order to displace purchased fuel. Such an arrangement is not only economical for both parties, but has the redeeming environmental advantage of reducing the quantity of wastes that must ultimately be disposed of by landfill or some other method. It should be noted that regardless of source, wastes must be destroyed safely, with the proper equipment in an environmentally sound manner.

ENERGY COST SUMMARY

The actual cost associated with each fuel should take into account the energy content of the fuel. To determine the incremental cost of energy for each fuel on an equal basis, the cost of each fuel should be compared in common units, which can be expressed as a cost per Btu (kWh), or a multiple thereof, where

1 kWh = 3413 Btu

Commonly used basis are kWh (1000 W), MWh (1,000,000 W), therms (100,000 Btu) and 10^6 Btu (1,000,000 Btu). This approach clearly illustrates that the cost of electricity per unit heat is much greater than that of gas or oil, making it prohibitively more expensive to heat with electricity in most applications. Another more subtle observation of a more detailed list is that residual oils are not only less expensive than lighter oils on a volume basis, but also have a higher higher heating value (HHV) as well. Table 6.1 summarizes the incremental costs presented above, which are used throughout the text.

Table 6.1
*Typical Incremental Energy Costs**

Energy Source	Cost (US$/Unit)	Energy Content	US$/$10^6$ Btu
Electricity			
Residential	0.09/kWh	3.413 Btu/W	26.37
Commercial	0.072/kWh	3.413 Btu/W	21.10
Industrial	0.0616/kWh	3.413 Btu/W	18.05
Natural Gas			
Residential	0.60/100 ft^3	1,000 Btu/ft^3	6.00
Commercial/Industrial	0.45/100 ft^3	1,000 Btu/ft^3	4.50
Fuel Oil	0.51/gallon	150,000 Btu/gal	3.40
Coal	000940.00/ton	13,000 Btu/lb	1.54
Waste	0.20/lb**	10,000 Btu/lb	20.00**

* Based on examples from the text. This data is fictitious and not representative of actual rates in any locale.
** Calculated value based on composition. The real value is the cost of the displaced fuel.

DISTRIBUTION LOSSES

The above costs represent primary utility billing costs that are measured as the utility enters the plant and, as such, do not take into account the distribution losses. For example, electrical distribution losses occur in the transformers and plant wiring. In a typical fuel oil system, the fuel oil is heated and constantly recirculated, both of which incur an economic penalty. However, once a system is in place and operating, the incremental cost associated with these factors can usually be ignored.

Plant Utilities

7

Plant utilities can be defined as those services that are distributed throughout the plant and are either consumed in the process or utilized in its support. Some of the more common plant utilities include water (potable, river, cooling, chilled), chilled brine, chilled glycol, air (compressed), industrial gases (nitrogen, oxygen, and the like), and steam.

The use of a plant utility is not without its costs, which are typically tabulated by a plant utilities engineer. In this section, typical plant utility costs based upon a representative plant utility distribution system will be calculated with the intent of illustrating the methodology necessary to calculate these costs.

Of extreme importance in the calculation of operating costs is determining the relationship between the system output and the primary utility input. It is this information that is essential to gaining a true perspective of operating costs, regardless of the process or piece of equipment.

EXTERNALLY POWERED UTILITY DISTRIBUTION

Externally powered plant utilities are those where no energy is required for distribution. Examples of this type are potable water, where the water is supplied to the plant under sufficient pressure for plant-wide distribution, and industrial gases, such as oxygen, nitrogen, or hydrogen, which are either compressed or delivered as a high pressure liquid and vaporized.

The total operating cost of the potable water system includes not only the purchase cost of the water, but also its distribution throughout the plant, including the costs associated with freeze protection and maintenance, which can easily add 25 percent or more to the cost, depending upon the geographic location and nature of the plant distribution system. The incremental cost of supplying one additional unit volume of potable water in the existing distribution system can be approximated by the cost of potable water.

Industrial gas costs will vary with location and quantity purchased. While maintenance costs associated with the total distribution system can be significant, the incremental costs of industrial gas usage can be approximated by the gas cost.

PUMPED DISTRIBUTION AND RECYCLE PUMPING SYSTEMS

Pumped distribution systems and recycle pumping systems typically contain a pump (and perhaps an additional booster pump) that is used to distribute the designated utility throughout the plant. Typical utilities with this distribution system include river water, cooling water, chilled water, chilled brine and chilled glycol systems.

Generation Costs

In addition to the cost of electricity, hidden costs can be associated with the generation of some plant utilities that may be difficult to fully quantify. Even river water, which is usually available gratis, is treated and pumped throughout the plant with freeze protection adding to steam, chemical, labor, and maintenance costs. Other utilities, such as cooling water and chilling systems, may have significant chemical, labor, and maintenance costs that can be quantified.

Distribution Costs

Adequate distribution to utility users dictates that the pumping system be capable of supplying adequate flow to any point on the distribution system regardless of whether system demand is high or low. Therefore, the distribution pumping system must have sufficient flow capacity and head to provide adequate flow anywhere in the distribution system at maximum flow conditions. As the pump is sized to handle this maximum load, it is not uncommon to install a large pump and operate it at 25 to 45 percent of capacity for over 95 percent of the time.

Assume that the distribution system above is designed to supply the plant with a maximum water flow of 250 gpm (946 lpm) at a minimum pressure of 80 psig (5.5 bar). As the filter system is designed to have a 15 psid (1 bar) pressure drop at maximum flow, the required pump capacity is 250 gpm at a discharge pressure of (80 + 15) psig × 2.31 feet/psi, or 219 feet (67 meters) of water column. A 7½ inch (191 mm) impeller will produce the desired head at the maximum flow at a pump efficiency of approximately 52 percent, utilizing a brake horsepower of approximately 22 hp (16.4 kW), such that a 25 hp (18.7 kW) motor would be applicable. The hydraulic energy produced by the system can be calculated as

$$(250 \text{ gpm} \times 80 \text{ psig} \times 2.31 \text{ feet/psi} \times 1.0/3960) \times 0.746 \text{ kW/hp}$$

or 8.70 kW. The billed electrical load can be estimated as

$$[\{(0.15 + 0.12) + (0.85 - 0.12) \times (22 \text{ hp}/25 \text{ hp})\}/(0.93 - 0.12)] \times 25 \text{ hp} \times 0.746 \text{ kW/hp}$$

or 21.0 kW. The overall efficiency is (8.70 kW/21.0 kW), or 41.4 percent. Should the system operate 98 percent of the time, the annual cost of operation is

$$(0.98 \times \text{US\$539.36/year} \times 21.0 \text{ kW})$$

or US$11,100/year. These costs can also be expressed relative to flow as (US$11,100/250 gpm), or US$44.40 per gpm per year.

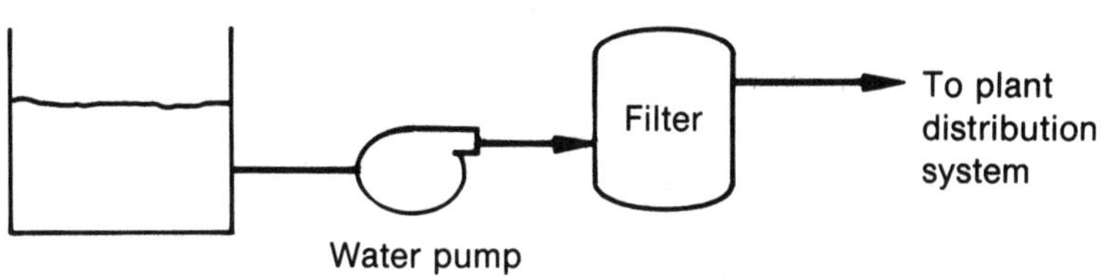

Figure 7.1
Plant Water System

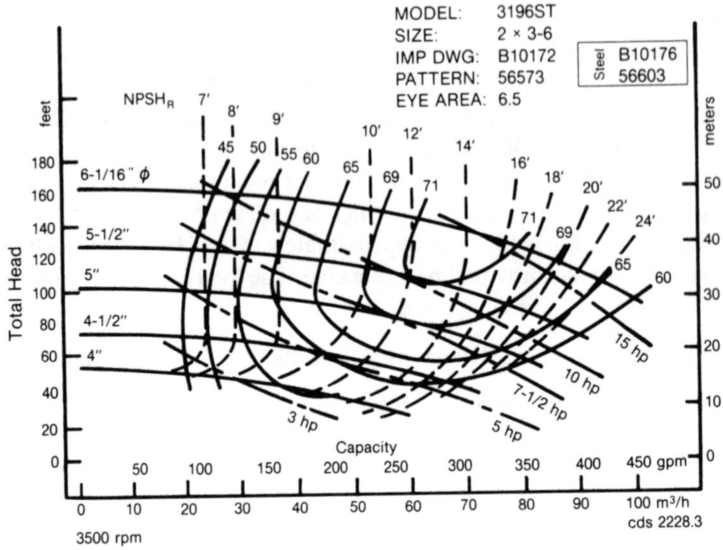

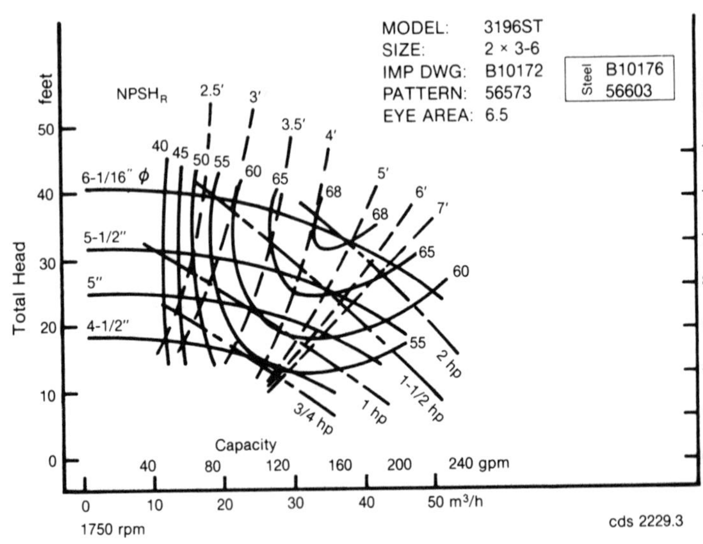

Figure 7.2
Centrifugal Pump Performance Curve

Assuming that the average system operating condition is 75 gpm (284 lpm), at which flow the discharge pressure is 115 psig (7.9 bar), the pump efficiency and brake horsepower at these conditions are approximately 35 percent and 14 hp (10.4 kW), respectively. The system pressure (after the filters) is

$$115 \text{ psig} - (75 \text{ gpm}/250 \text{ gpm})^2 \times 15 \text{ psid}$$

or 113.7 psig (7.8 bar).

The billed electrical load of the system can be estimated as

$$[\{(0.15 + 0.12) + (0.85 - 0.12) \times (14 \text{ hp}/25 \text{ hp})\}/(0.93 - 0.12)] \times 25 \text{ hp} \times 0.746 \text{ kW/hp}$$

or 15.6 kW. Assuming the system is operating 98 percent of the time at 75 gpm, the annual operating cost associated with the motor can be estimated as

$$(0.98 \times \text{US\$539.36/year} \times 15.6 \text{ kW})$$

or US$8246/year. The actual annual operating cost can also be related to flow as US$8246/75 gpm actual flow, or US$109.94/gpm. The overall efficiency of the system can be calculated as

$$(75 \text{ gpm} \times 113.7 \text{ psig} \times 2.31 \text{ feet/psi} \times 1.0 \times 0.746 \text{ kW/hp}/3960)/15.6 \text{ kW}$$

or 23.8 percent. If one uses the 80 psig system requirement in the calculations to provide a truer measure of system efficiency, the system efficiency decreases to

$$(80 \text{ psig}/113.7 \text{ psig}) \times 23.8 \text{ percent}$$

or 16.7 percent.

As can be seen from the above analysis, the electrical operating costs are significant, and, although there is a reduction in energy costs when a pump is throttled, the hydraulic efficiency at low flows can decrease dramatically from that at design operating conditions. Of extreme importance in the above analysis is the concept of examining the cost of the final product as a function of billed costs.

While a reduction in the consumption of a purchased plant utility directly results in cost savings, a reduction of consumption of a pumped utility generally results in lower operating costs and reduced hydraulic system efficiency. As a result, cost reductions of pumped systems are not directly proportional to the reduction in consumption, and the percentage cost reduction can be significantly less than the percentage reduction of load.

PLANT AIR COSTS

Plant air costs are dependent upon the efficiency with which the air is compressed, the nature of the drying equipment, and the plant distribution system. Of particular concern in the determination of plant air cost is the relationship between the hydraulic energy contained in the compressed air leaving the system and the utilities consumed to produce the compressed air.

Air compressors consume approximately 1 bhp (0.75 kW) of mechanical energy to produce 4 cfm (0.113 m^3) of air at a nominal 100 psig (6.9 bar). While some compressors may do slightly better or worse than 4 cfm per bhp, the operating cost, maintenance cost, and reliability of different types of compressors can be very different.

Consider a positive displacement compressor where a valve controls the discharge pressure by bleeding air to the atmosphere. This control is rather straightforward, but results in fully loaded compressor operation and billing, independent of load. For example, a fully loaded 100 hp (75 kW) motor operating a compressor that produces 400 cfm (11.3 m^3/min) of air will use

$$[0.18 + 0.82 \times (75/75)] \times 75 \text{ kW}/0.90$$

or 83 kW. The compressor will continue to draw appoximately 83 kW when the load is diminished as unneeded compressed air is vented to atmosphere.

Other compressors have mechanisms that allow the compressor to operate unloaded under low load conditions and produce no compressed air. In this mode of operation, the mechanism is loaded when the compressed air pressure falls below a preset pressure and unloads when the pressure rises above a higher preset pressure. The nature of this control causes the plant air pressure to oscillate between the preset pressures and also reduces electrical consumption.

Electrical consumption when a similar compressor is fully loaded will be the same as that described above. However, when it is unloaded, the electrical consumption can be estimated as

$$[0.18 + 0.82 \times (7.5/75)] \times 75 \text{ kW}/0.90$$

or 22 kW, assuming that 10 percent of the motor capacity is required to operate the compressor while it is unloaded. Electrical consumption at 70 percent load can be estimated as $(0.7 \times 83 \text{ kW}) + (0.3 \times 22 \text{ kW})$, or 65 kW. Power consumption is less than that of the above example; however, the compressor operates unloaded for 30 percent of the time, during which 22 kW are consumed to produce no compressed air.

Assuming that the unloading compressor remains loaded during a 15-minute demand period at some time during the billing cycle, the demand charges associated with both of the above compressors will be 83 kW.

The losses associated with the transformer and distribution systems are typically neglected but can be estimated at approximately 1.5 percent of the average current draw of the motor. In the case of the fully loaded compressor, the current drawn is the full load amps (FLA) of the motor, whereas the current drawn when unloaded can be estimated as

$$[(0.35) + (0.65 \times 10/100)] \times \text{FLA}$$

or 0.415 FLA. The average current draw of this compressor is

$$[(0.7) + (0.3 \times 0.415)] \text{ FLA}$$

or 0.82 FLA. From the above calculations for current draw, more capacity, and line voltage information, transformer losses can be calculated.

Air dryers typically introduce energy and capacity losses to the compressed air system due to pressure drop and regeneration operations. Of importance in determining the cost of compressed air is the operating cost of the air dryer, which takes into account electricity, compressed air, and any other utilities required. When calculating the efficiency of the compressed air generation system, the output conditions of compressed air pressure and flow should be measured at the point of discharge to the plant, downstream of the air dryer.

The operating cost associated with the fully loaded compressor and an air dryer that has its 10 kW heaters on 50 percent of the time can be calculated as

$$(83 \text{ kW} + 0.5 \times 10 \text{ kW}) \times \text{US\$539.36/kW·yr})$$

or US\$47,464. The operating cost associated with the unloading compressor is

$$(65 \text{ kW} + 0.5 \times 10 \text{ kW}) \times \text{US\$539.36/kW·yr}$$

or US\$37,545. Transformer losses associated with a fully loaded compressor (which illustrates the maximum losses) is approximately

$$[0.015 \times (75 \text{ kW}/0.9 \text{ efficiency})]$$

or 1.25 kW, which results in a US\$674 economic loss. Note that the transformer losses are small and typically neglected.

The cost of compressed air for each compressor can be calculated at 70 percent of capacity as

$$\frac{\left(\frac{\text{US\$47,464}}{\text{yr}}\right)\left(\frac{\text{yr}}{8760\ \text{hr}}\right)\left(\frac{\text{hr}}{60\ \text{min}}\right)}{(0.7 \times 400\ \text{ft}^3/\text{min})} = \text{US\$0.32/1000 ft}^3$$

for the fully loaded compressor, and

$$(\text{US\$37,545/US\$47,464}) \times \text{US\$0.32/1000 ft}^3$$

or US\$0.25 per 1000 ft^3 (28.3 m^3) for the unloading compressor. The incremental cost of compressed air produced from the loaded compressor is US\$0.00, as load does not affect power consumption, whereas the incremental cost of compressed air produced by the unloading compressor can be estimated by using the above equations to estimate the operating costs at two load conditions.

STEAM COSTS

The cost of plant steam production includes not only the cost of fuel but also the cost of electricity, labor, and raw materials required to provide the utilities required for steam generation. The primary utilities include feedwater treatment systems, steam and condensate chemical programs, fuel distribution systems, blowdown systems, condensate return systems, and the like. The incremental cost of steam production can be reasonably approximated by

Fuel cost per unit energy content/Boiler efficiency

in appropriate units.

For example, consider a boiler with an average combustion efficiency of 75 percent, 10 percent blowdown, and a nominal 2 percent radiated heat losses, fired on purchased natural gas, which costs US\$4.50 per 10^6 Btu (US\$15.36 per MW).

Assuming that the boiler nominally produces 130 psig (9 bar) saturated steam at a temperature of 356°F (180°C) and make-up water at 50°F (10°C) and condensate are introduced to the boiler system at an average temperature of 176°F (80°C), the heat required to evaporate the water is

$$[(356°\text{F} - 176°\text{F}) \times 1.0\ \text{Btu/lb°F}] + 866\ \text{Btu/lb}$$

or 1046 Btu per pound (7.872 MW per metric ton).

The heat loss due to the blowdown, which is 10 percent of the steaming rate, is

$$[0.10 \times (356°F - 50°F) \times 1.0\ Btu/lb°F]$$

or 31 Btu per pound (0.209 MW per metric ton).

The incremental cost of steam can be estimated as

$$(US\$4.50/10^6\ Btu) \times (1000 - 31\ Btu/lb) \times 1000\ lbs/[0.75 \times (1 - 0.02)]$$

or US$5.93 per 1000 pounds (12.86 per metric ton). The true incremental cost of steam will be slightly higher due to incremental increases of such items as chemical treatment, electrical energy consumption, and the like that have not been considered in the above calculations.

Application of Control Valves

8

Control valves are used extensively throughout industry as final control elements to manipulate the process by varying a restriction in a flowing fluid, thereby throttling fluid flow. In essence, a *control valve* is a device that dissipates hydraulic energy in a controlled manner. The control valve is dominant among final controlling elements and, barring unforeseen technological advances, is expected to maintain its prominent position for many years to come.

The following discussion represents an insight into the application of control valves and an appreciation of control valve sizing, selection, and installation considerations.

CONTROL VALVE COMPONENTS

Control valves generally consist of a valve body and a valve actuator. The *valve body* is the portion of the valve across which the pressure drop is taken and contains the valve's working components that constitute the restriction to flow. The energy dissipated by the control valve can be calculated as

$$\text{hp} = \frac{W_{\text{lb/min}} \times \Delta P_{\text{psi}}}{14288}$$

$$\text{kW} = \frac{W_{\text{kg/min}} \times \Delta P_{\text{bar}}}{599.8}$$

or in a more useful form for liquids,

$$\text{hp} = Q_{\text{gpm}} \times \Delta P_{\text{ft}} \times \text{SG}/3960$$

$$\text{kW} = Q_{\text{lpm}} \times \Delta P_{\text{ft}} \times \text{SG}/4601$$

where SG is the specific gravity of the liquid and 1 hp = 0.746 kW.

Parts within the valve body are in contact with the fluid and are termed *wetted parts*. Regardless of the service, all wetted parts must be compatible with the fluid at the operating conditions that exist within the valve for good service life and a minimum of leaks through packing and/or seals. Phenomena such as flashing, cavitation, two-phase flow, abrasion, and the like, must be dealt with in many applications.

The *valve actuator* is the portion of the control valve that supplies the force necessary to operate the valve. Sufficient force must be supplied by the actuator to enable control valve operation at all valve positions under extreme operating conditions. Non-wetted components must be compatible with the environment (for example, prevailing weather conditions when installed outdoors) in order to function reliably.

Control valve designs are classified by the motion of valve movement. Some control valve designs are linear in nature, while other designs are rotary and are often referred to as quarter-turn designs.

TYPICAL CONTROL VALVE DESIGNS

Control valve designs utilize many different constructions to satisfy the many and varied applications present in industry. Different types of control valves are presented below with the intent of giving the reader a cursory insight into typical control valve construction and applications. The control valves described are applicable to corrosive and non-corrosive liquids and gases, unless otherwise noted.

Globe Valves

The term *globe valve* is used to designate any of a number of valve varieties. The common feature among globe valves is a plug, attached to the valve stem, which moves linearly so as to vary the valve opening, thereby restricting fluid flow.

There are many globe valve designs, the most common being the single-port globe valve shown in Figure 8.1a. Applications of the single-port valve

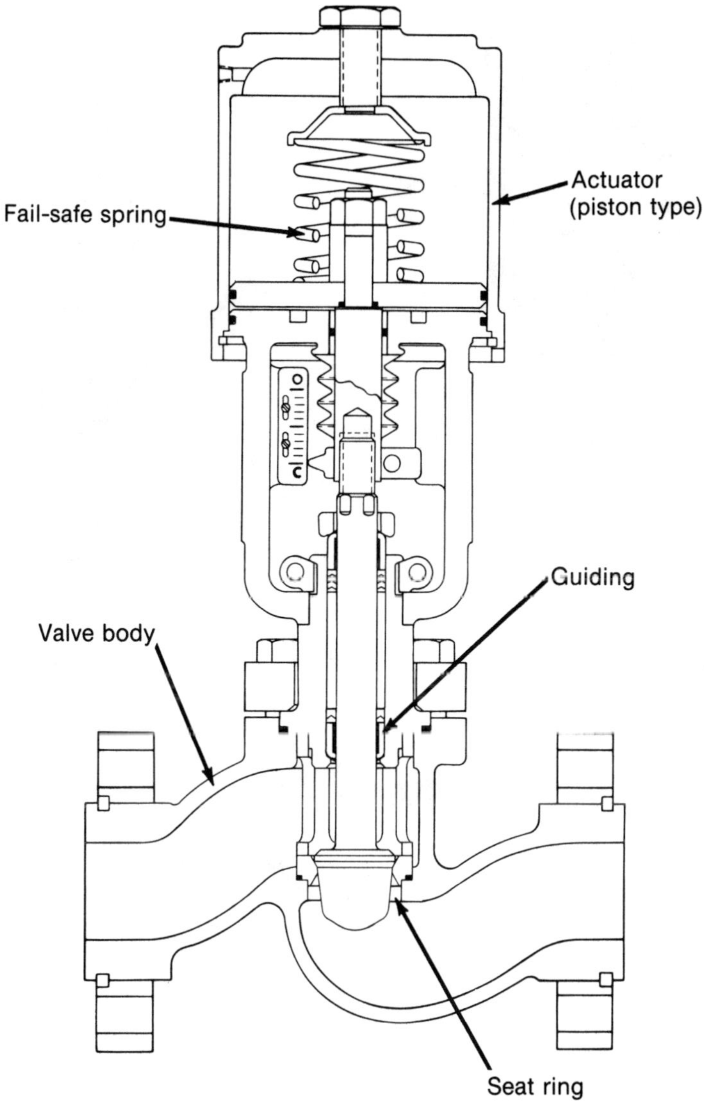

(a) Globe Valve with Piston-Type Actuator

(courtesy Valtek)

Figure 8.1
Globe Valves

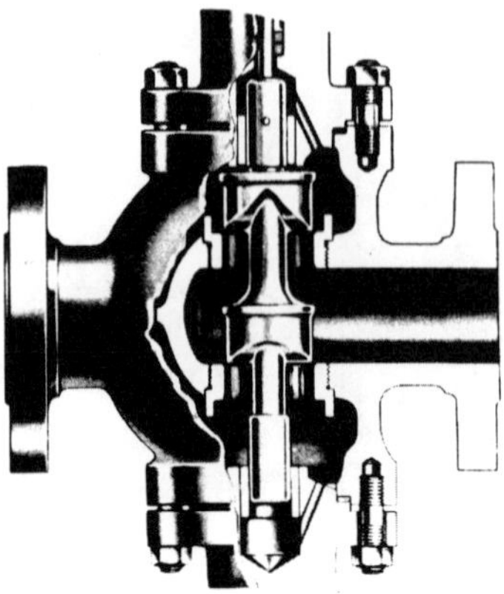

(b) Double Port Globe Valve

(courtesy Masoneilan International, Inc.)

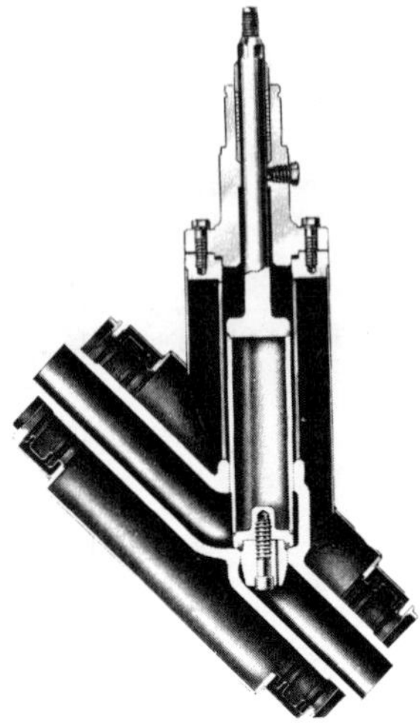

(c) Y-Pattern Valve

(courtesy Masoneilan International, Inc.)

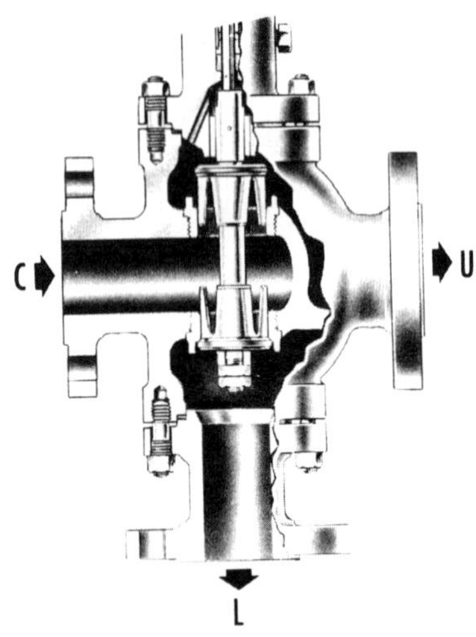

(d) Three-Way Valve-Diverting Service

(courtesy Masoneilan International, Inc.)

Figure 8.1
(continued)

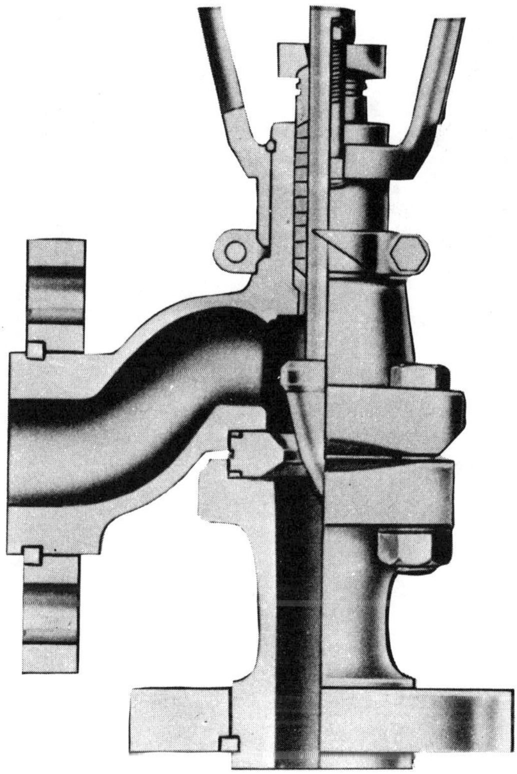

(e) Angle-Type Split Body

(courtesy Masoneilan International, Inc.)

Figure 8.1 (continued)

vary and include highly corrosive chemicals provided that suitable materials of construction are used. Other globe body designs are applied in special applications such as in high differential pressure drop applications, three-way valve applications, and the like.

Gate Valves

A *gate valve* (see Figure 8.2) is linearly actuated and utilizes a flat plate that is slid into the flow stream, the position of the gate controlling the amount of restriction of the fluid. These valves are typically used for on/off service.

(a) Multi-Orifice Gate Valve
(courtesy Jordan Valve Division, Richards Industries, Inc.)

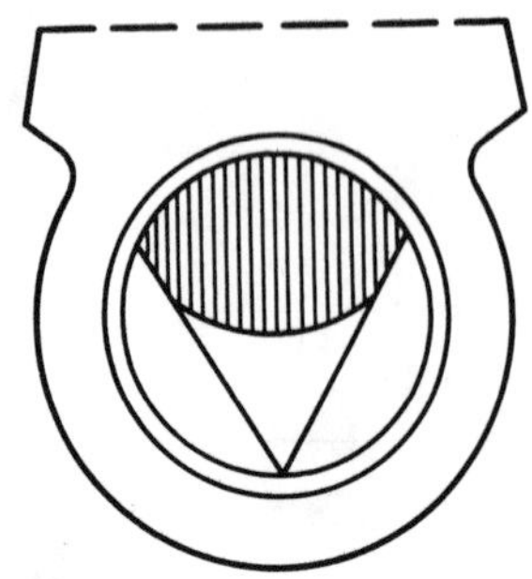

(b) V-Orifice Gate

Figure 8.2
Gate Valves

Diaphragm Valves and Pinch Valves

Diaphragm valves are linear and utilize a flexible diaphragm to restrict fluid flow (see Figure 8.3). These valves are commonly used on "difficult" fluids (such as corrosive liquids or slurries) due to the availability of corrosion resistant liners and diaphragms.

Pinch valves are a variation of the hose clamp and are utilized in various applications, including slurries (see Figure 8.3). As high wear can occur where the liner is bent 180° due to high fluid velocity when the valve is nearly closed, annular control valves, which maintain a circular flow pattern but do not shut off completely, are often applied in throttling applications.

Butterfly Valves

Butterfly valves utilize rotary motion to position the valve to obtain the desired restriction to fluid flow. The butterfly type is the most commonly applied rotary valve and a large number of design variations exist (see Figure 8.4). Applications include highly corrosive chemicals due to the availability of corrosion resistant metals and liners.

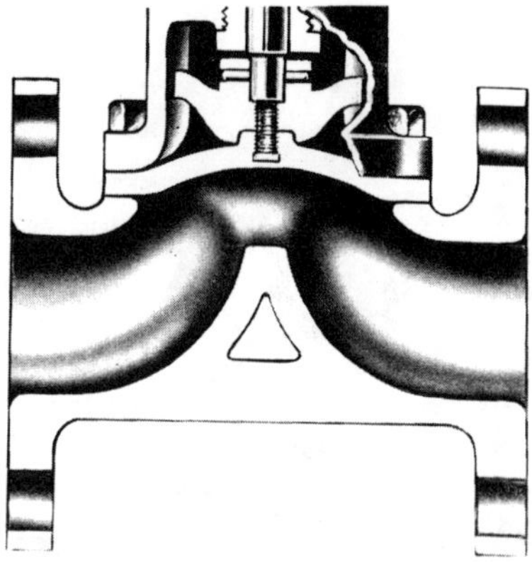

(a) Diaphragm Valve

(courtesy Masoneilan International, Inc.)

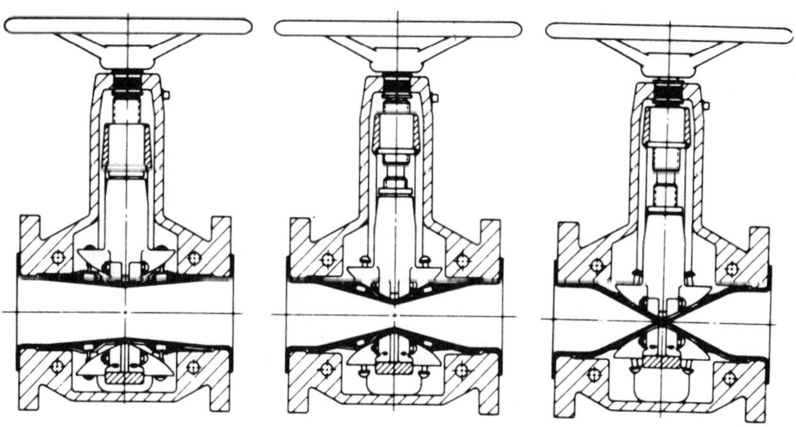

(b) Pinch Valve

(courtesy Resistoflex Corporation)

Figure 8.3

Diaphragm and Pinch Valves

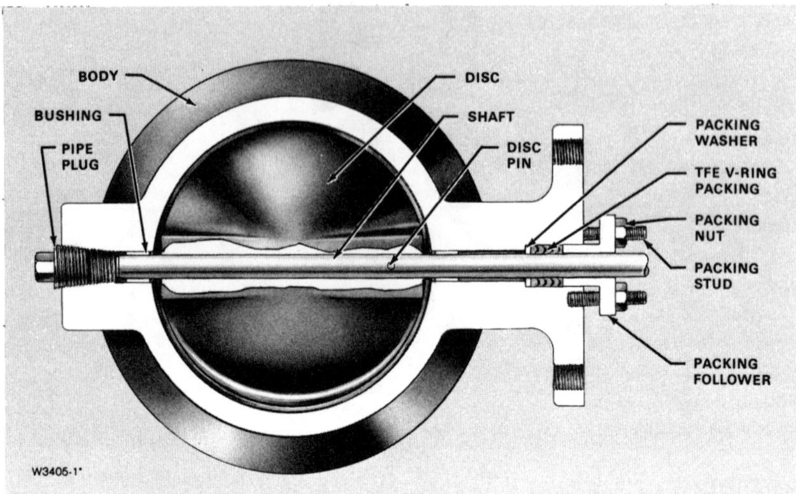

(a) Partial Sectional of Butterfly Valve

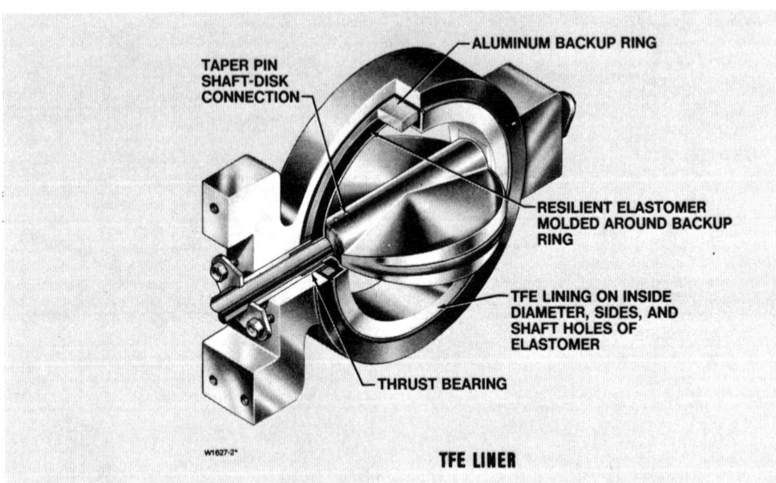

(b) Lined Butterfly Valve

Figure 8.4
Butterfly Valves

(courtesy Fisher Controls)

(c) Eccentric Disk Valve Body with Diaphragm Actuator and Positioner

Figure 8.4 (continued)
(courtesy Fisher Controls)

Ball Valves

A *ball valve* is a rotary valve wherein the hole in a sphere is rotated in the fluid stream to vary the restriction to fluid flow (see Figure 8.5). Many designs are available and some utilize segmented balls instead of the full ball design. The virtually obstructionless flow path in the fully open position allows their use on applications with some solids content when suitable materials of construction are used.

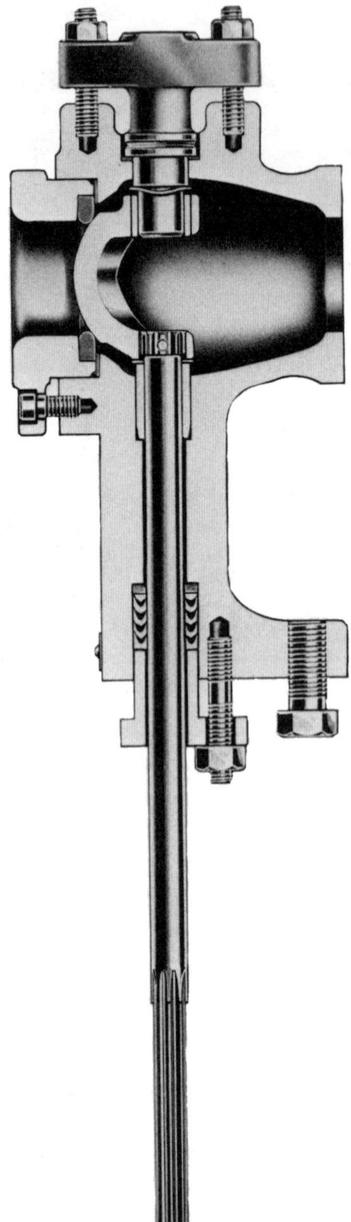

Figure 8.5
Ball Valves

(a) Sectional Typical of 2 through 12-inch Vee-Ball® Valve
(courtesy Fisher Controls)

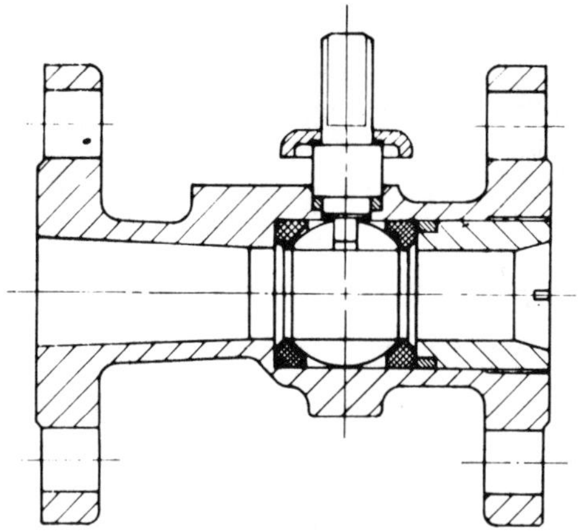

(b) Full Ball Valve

(courtesy Jamesbury Corporation)

(c) Vee-Ball® Valve with Actuator and Positioner

(courtesy Fisher Controls)

Figure 8.5
(continued)

Plug Valves

The *plug valve* is a rotary valve where a conical or cylindrical plug is rotated to control the flow passage and hence the restriction to flow (see Figure 8.6). Application of these valves include highly corrosive chemicals due to the availability of suitable metal construction and liners.

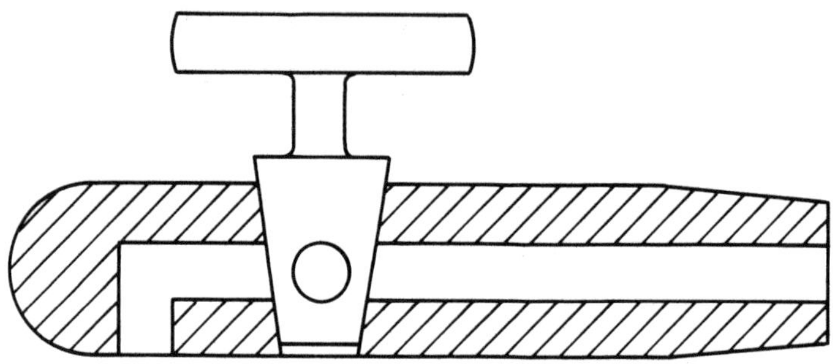

(a) An Early Plug Valve

(b) Eccentric Cylindrical Plug

(courtesy Dezurik Corp.)

Figure 8.6
Plug Valves

(c) Eccentric Spherical Plug
(courtesy Masoneilan International, Inc.)

Figure 8.6 (continued)

Miscellaneous Valve Designs

A number of other valve designs, such as multi-ported valves, dampers, and vanes, are available for special as well as general applications. A 12-bit digital control valve is shown in Figure 8.7.

CONTROL VALVE SIZING

The generally accepted method of *sizing* control valves involves calculating the required valve capacity, then selecting a valve that is capable of passing the required capacity at a reasonable valve position (typically 30–75 percent of valve travel, depending on valve design).

C_v, which is a measure of valve capacity, is defined as the amount of water in gallons per minute that will flow through the valve at 60°F (15.6°C) with a pressure drop of 1 pound per square inch differential (psid). The relationship between C_v, fluid flow, and pressure drop in a simple liquid system is given by

$$C_v = Q_{\text{gpm}} \times \sqrt{\frac{\text{SG}}{\Delta P_{\text{psi}}}} = \frac{0.0696\ \text{W}_{\text{kg/min}}}{\sqrt{\rho_{\text{g/cm}^3} \times \Delta P_{\text{bar}}}}$$

where P is the pressure drop across the valve. Different equations and techniques are used to calculate C_v for more demanding applications, such as cavitating fluids, flashing fluids, viscous fluids, two-phase fluids, and the like.

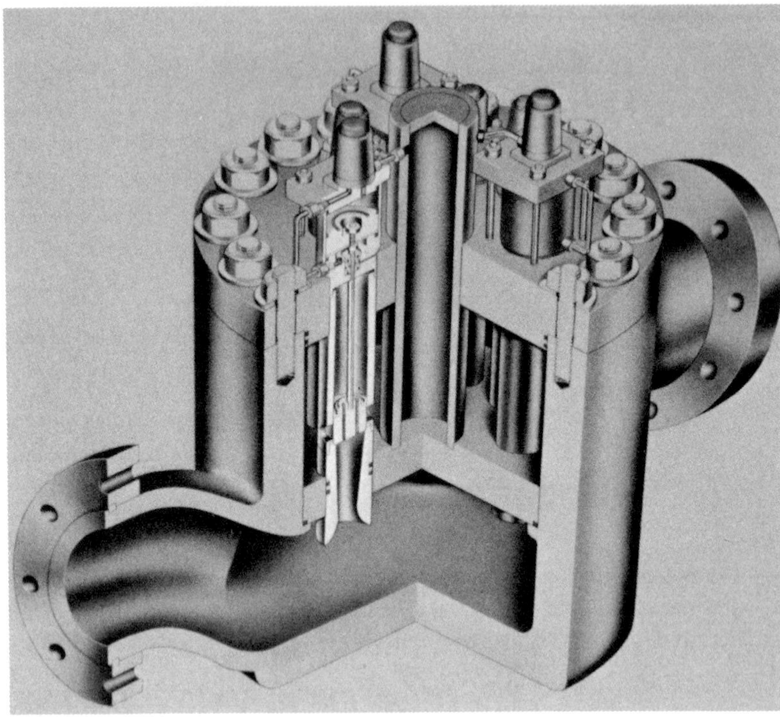

Figure 8.7
Digital Valve (12-bit)
(courtesy Digital Valve Company)

CONTROL VALVE CHARACTERISTICS

The relationship between the position of the control valve and its capacity is termed the *inherent flow characteristic* of the valve. The capacity of a control valve with linear trim increases linearly with valve travel. The capacity of a control valve with equal percentage trim increases exponentially with valve travel, that is, equal increments of valve travel result in equal percentage changes in capacity (see Figure 8.8). Other modified characteristics are also available.

It should be noted that due to the characteristics of the system in which a control valve is installed, the installed flow characteristic can vary significantly from the inherent flow characteristic of the valve. The nature of the

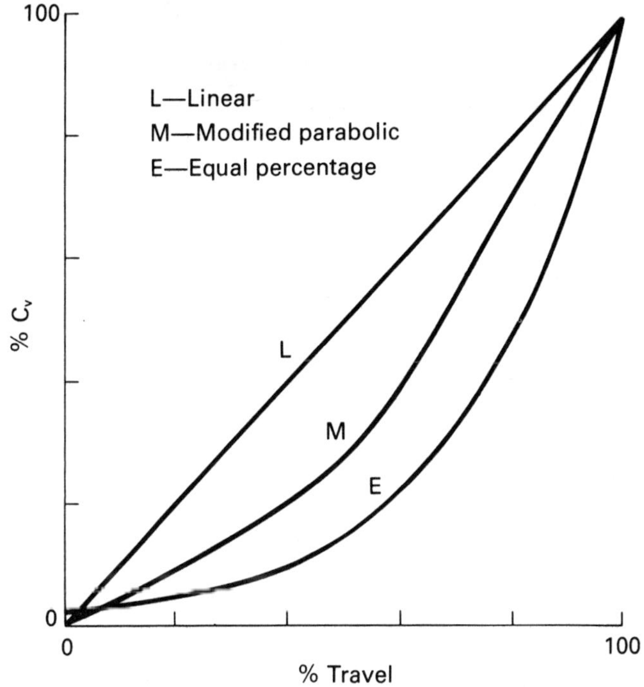

Figure 8.8
Ideal Flow Characteristics

measurement device and the characteristics of the valve and actuator may cause a control loop to become unstable under certain conditions, resulting in rangeability limitations.

CONTROL VALVE ACTUATORS

Control valve *actuators* are those devices that provide the motive force to operate the control valve (see Figure 8.9). In general, these devices utilize a discrete energy source with which to provide the forces necessary for valve movement. A loss of energy source may cause the control valve to fail open, fail closed, fail in last position, or fail to an intermediate position, depending upon the design and configuration of the actuator and the control valve.

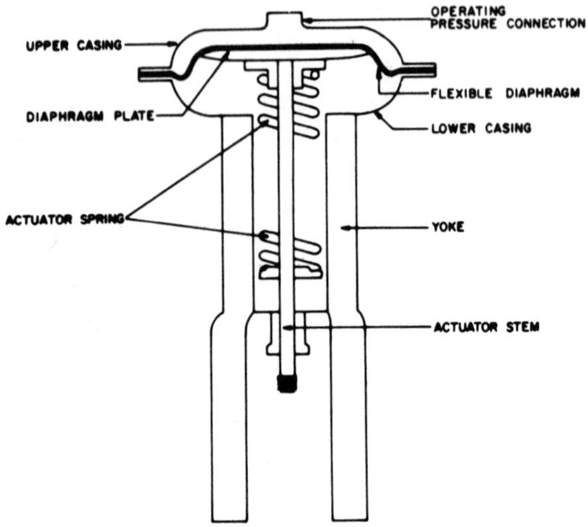

(a) Spring and Diaphragm Actuator

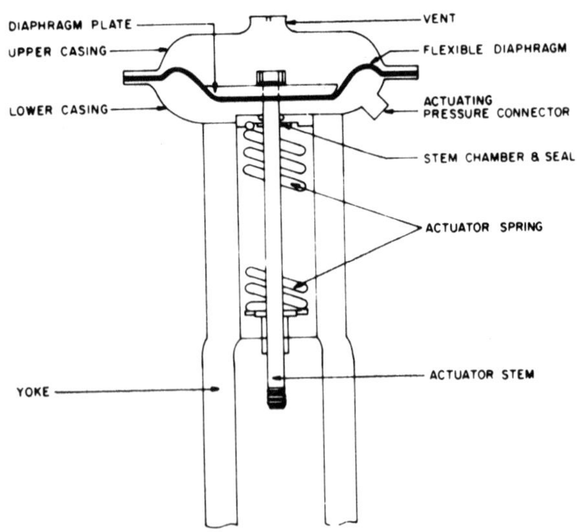

(b) Reverse Acting Actuator

Figure 8.9
Actuators

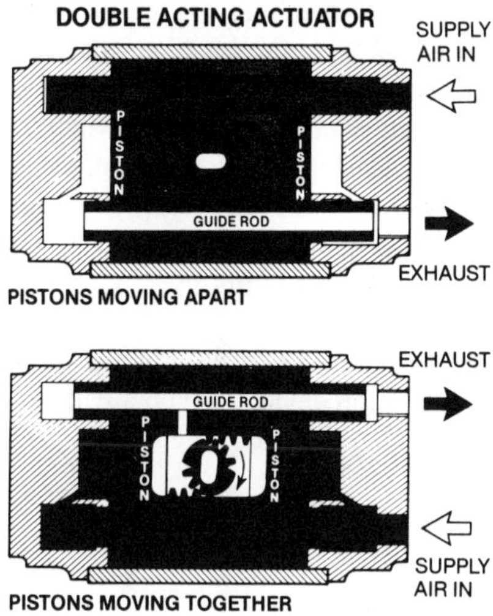

(c) Double Acting Actuator

(courtesy Worcester Controls)

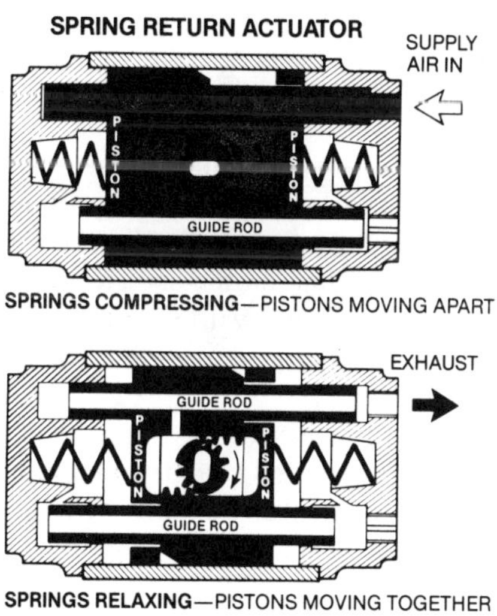

(d) Spring Return Actuator

(courtesy Worcester Controls)

(e) Electric Valve Actuator with Integral Positioner

(courtesy Harold Beck and Son)

Figure 8.9 (continued)

The most commonly used energy source for industrial applications is compressed air. *Pneumatic* actuators, which operate on compressed gases, are simple and easily repaired because of their rugged, reliable design. *Electric* actuators are used in many applications where compressed gas is not available or where electrical operation is desireable, such as for continued operation of a control valve should the compressed gas system fail. Hydraulic actuators are used to generate sufficient force to operate extremely large valves, while self-contained regulators use the motive force of the fluid to operate the actuator.

Control valve actuators are generally divided by their nature of motion into linear and rotary motion devices, which are applied to linear and rotary valves, respectively.

CONTROL VALVE ACCESSORIES

Accessories are often required to enable the control valve and actuator to operate in the instrumentation system. Typically, an electronic signal is provided from the instrumentation system and converted into a pneumatic signal compatible with the actuator. For example, a *current-to-pneumatic converter* (I/P) is required to convert the 4–20 milliamp (mA) electronic signal into a 3–15 psig (0.2–1.0 bar) pneumatic signal to operate the actuator. A *pressure regulator* is often used to control and limit compressed air pressure to the actuator and accessories.

The application may require a *positioner* that modulates the energy source to control the opening of the control valve so that it is proportional to the signal from the instrumentation system. Positioners may be provided internally or externally to the actuator, depending on the actuator design (see Figure 8.10). It should be noted that an actuator without a positioner can be highly nonlinear and that the control valve position will not necessarily correspond to the instrumentation signal.

Limit switches may be required to sense open, closed, or intermediate valve positions, while a *solenoid valve* may be required in some applications, such as on/off operation or where a preset control valve position is desired. Manual *handwheels* and *limit stops* are also available on many control valve actuators.

CONTROL VALVE INSTALLATION

A source of motive force must be distributed to each control valve, unless the control valve is self-actuated. For pneumatic valves, compressed gas must be piped to each control valve actuator, while electric actuators require wiring an electrical power circuit to each control valve. Depending on the location of the hydraulic actuator, electrical power and/or hydraulic piping may be required to each hydraulically operated actuator.

Linear control valves are typically mounted with the actuator vertical to minimize stresses on the assembly as well as to minimize the effects of rain or snow in outdoor applications. Rotary control valves are usually mounted with the valve shaft in the horizontal plane to allow dirt in the pipe to pass freely below the valve shaft. Control valves should be located in places that are accessible for maintenance, provided that system performance does not suffer.

Some applications, such as where a high-capacity control valve located downstream of elbows, tees, valves, pumps, and blowers, may require

(a) Rotary Control Valve Body with Diaphragm Actuator and Positioner

(courtesy Fisher Controls)

Figure 8.10
Actuators and Positioners

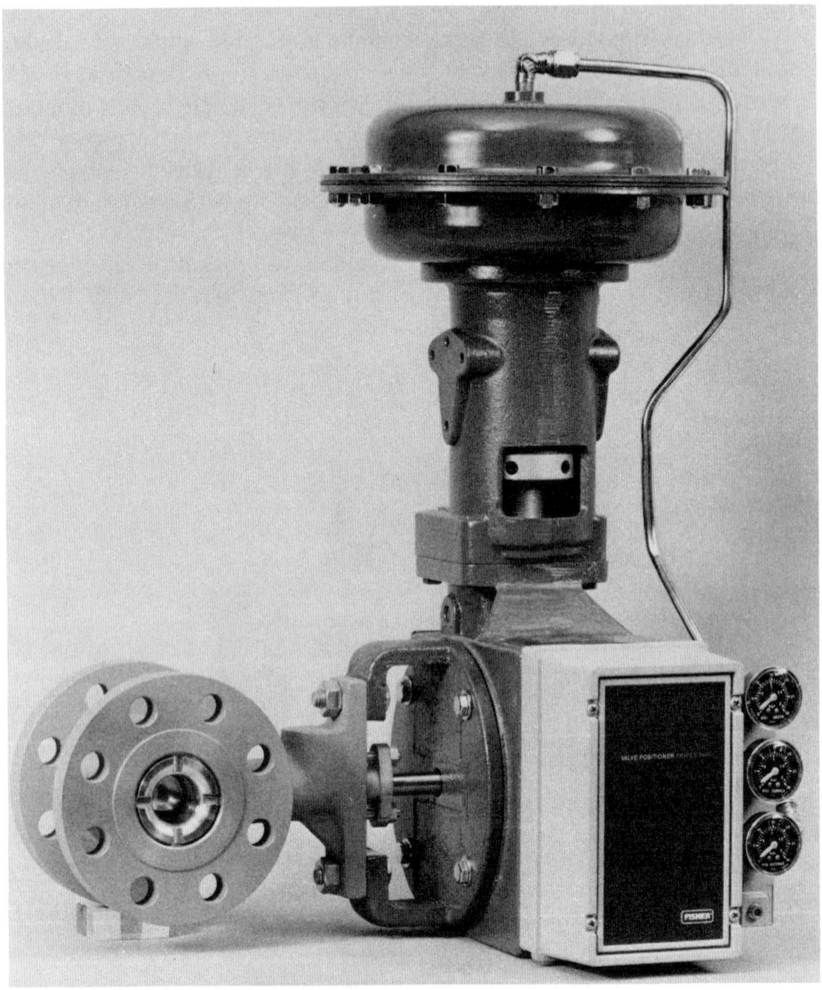

(b) Linear Control Valve with Piston Actuator and Positioner

(courtesy Fisher Controls)

Figure 8.10
(continued)

straight run upstream of the inlet reducer to the control valve to avoid hydraulic effects on control valve performance.

Control valves that are installed in continuously operating processes often utilize *block* and *bypass* valves to allow removal of the control valve for repair (see Figure 8.11). Block and bypass systems are expensive and can fail to seal properly when needed. Block and bypass systems should be employed only where required.

Fluid phenomena can affect valve performance in some applications. High fluid velocity may cause *flashing* in an upstream reducer and *two-phase* flow conditions upstream of the control valve, adversely affecting control valve performance. Flashing across a control valve with insufficiently sized discharge piping can choke the flow and reduce the system capacity.

CONTROL VALVE SELECTION

Control valve *selection* encompasses consideration of many diverse attributes that include, but are not limited to, pressure-temperature constraints, limitations of shut-off pressure differentials, operating temperature limitations, inherent flow characteristics, installed flow characteristics, sizing and

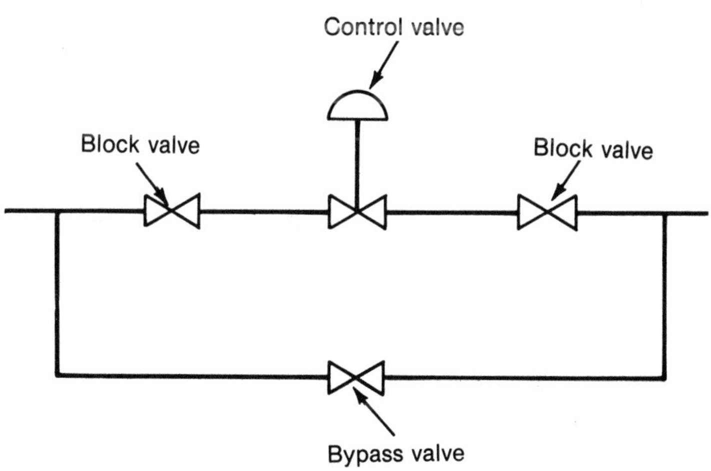

Figure 8.11
Block and Bypass Valving

rangeability, linearity, seat leakage, end connections, cavitation, cost, maintenance, noise considerations, body materials of construction, trim materials of construction, applicability of elastomer and gasketing materials, instrumentation system requirements, and installation constraints.

As can be seen from the above list, the number of considerations are many; however, in the majority of control valve applications, only a few of these criteria will be considered in detail in the selection process of an acceptable control valve. While experience is a valuable guide to control valve selection, control valve standards, practical industrial texts, and manufacturers' literature are good sources of practical control valve application information.

DAMPERS

Dampers are commonly applied to centrifugal fan applications to throttle flow. In gas applications, the damper may be located upstream or downstream of the fan. Power consumption associated with the throttling of the fan inlet is typically 5–20 percent less than that of throttling the fan outlet, depending on load. Inlet guide *vanes* are used to reduce fan capacity as well as provide rotation of the gas upstream of the impellor, reducing entrance losses.

Alternate Final Control Elements

9

Final control elements other than control valves have many appropriate applications throughout industry, which in some applications compete with control valve technology. Alternate final control element technologies often offer superior performance with reduced installed, operating, and maintenance costs. The applicability of alternate final control elements should be based upon process requirements, reliability, and economics.

Alternate final control elements fall into one of two general categories: variable capacity, where control is integrated into mechanical equipment, and variable speed drives, where the speed of the rotating equipment is varied. Both technologies have evolved over the course of time and have generally not competed with each other, as equipment dictated by size, process requirements, and industry practice was usually purchased as a package system. This often results in a significant loss of economic opportunities for the user due to limited ability on the part of the user and/or manufacturer to mix and match the best process equipment with the best available control equipment.

A direct result of speed control is a reduction in the brake horsepower at reduced speeds with resultant electrical savings. In most cases, reduced speed will result in reduced wear and tear on the equipment with a corresponding reduction in maintenance. Depending on the design, capacity controls may or may not result in significant electrical cost savings.

The following sections describe viable techniques utilized to throttle equipment to develop a working knowledge of the operating principles and attributes of various alternate final control element designs from a user's perspective.

VARIABLE CAPACITY EQUIPMENT

Manufacturers often design mechanical controls as an integral part of the equipment, the primary purpose of which is to control equipment in a desired manner. A secondary and often overlooked result of these controls is to reduce the energy waste associated with throttling the discharge of the equipment.

Applications of variable capacity equipment include:

(1) Positive displacement metering pumps whose displacement can be altered while operating, thereby adjusting liquid flow.
(2) Fans and blowers with inlet guide vanes that can be positioned while the equipment is operating, thereby adjusting gas flow.
(3) Reciprocating positive displacement compressors that have unloaders that prevent one or more cylinders from doing work, thereby adjusting capacity (in discrete steps).
(4) Axial positive displacement compressors where the compression ratio and/or capacity can be regulated while the equipment is operating.
(5) Cooling tower fans with blades, the pitch of which can be varied while the equipment is operating to vary cooling capacity.

The techniques with which the above concepts are applied to mechanical equipment are usually proprietary, and their effectiveness will vary. As with other final control elements, reliability and control system stability can pose problems when the equipment is operated in a plant environment.

MECHANICAL TRANSMISSION

In popular use, *mechanical transmissions* are rugged, easily understood devices that vary the speed of a piece of rotating equipment while the motor rotates at full speed. While adjustable pulleys are often used, other designs such as belts or chains may be used to effectively alter speed. Often, equipment is supplied with more than one pulley in order to adjust the speed in discrete steps, in some designs, while the equipment is rotating (see Figure 9.1).

This type of drive is available for fractional horsepower motors through tens of thousands of horsepower. Speed is generally adjustable over a 10 to 1 range.

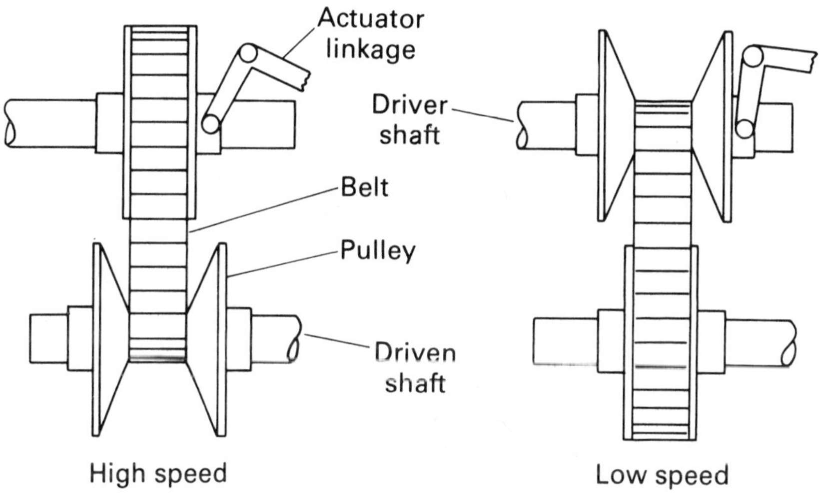

(a) Pulley Type

(b) Pulley Type with Remotely Controlled Actuator

(courtesy Hi-Lo Manufacturing Co.)

Figure 9.1
Mechanical Variable Speed Drives

(c) Beister Dry Traction Mechanical Adjustable Speed Drive

(courtesy Sumitomo Mach. Corp. of America)

Figure 9.1 (continued)

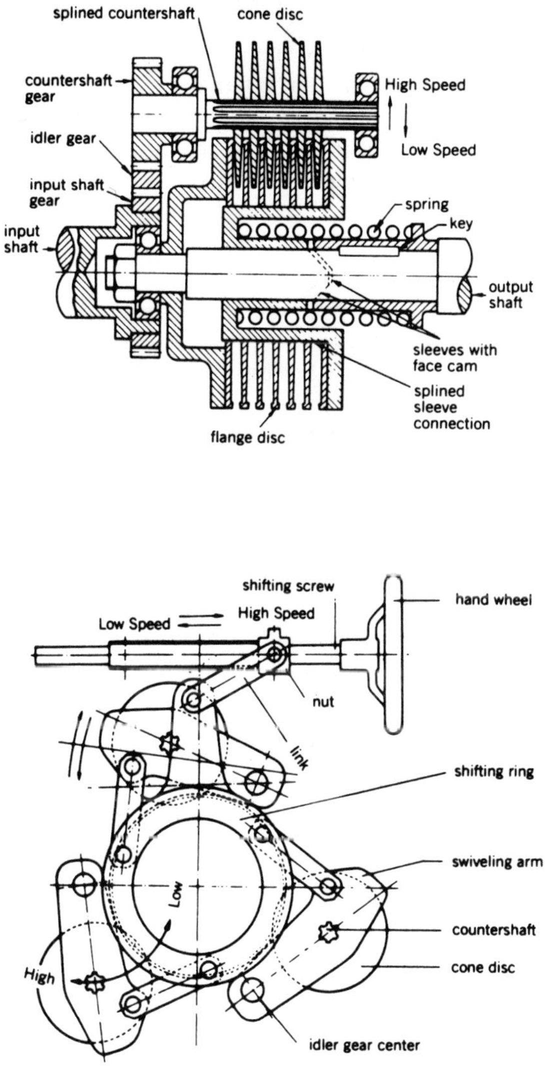

(d) Beier Cyclo Mechanical Adjustable Speed Drive

(courtesy Sumitomo Mach. Corp. of America)

Figure 9.1 (continued)

FLUID-POWERED DRIVES

Fluids that have sufficient pressure energy can be used to rotate a piece of equipment. These drives utilize an external power source that imparts energy to the fluid and a prime mover. *Fluid-powered drives* typically maintain a constant efficiency over most of the upper portion of their operating ranges. Therefore, fluid requirements tend to approximate load over this range.

Pneumatic Drives

Pumps and other small rotating pieces of equipment can be rotated using air motors. The speed of the air motor can be varied by adjusting a control valve, which throttles the amount of compressed air to the motor. These motors are typically applied only to small equipment due to the high cost of producing compressed air (see Figure 9.2).

Applications of pneumatic control of variable speed equipment include:

Small pneumatic hand tools
Boiler and cooling tower chemical feed pumps
Viscous liquid pumping applications
Laboratory pumps

Steam Turbines

Steam turbines are commonly used to control the speed of rotating equipment where suitable quality steam is available. Turbine speed is varied by throttling the steam supply with a control valve, which may be adjusted by the instrumentation system manually or, more commonly, by a steam regulator that controls a desired process parameter such as the discharge pressure of the equipment (see Figure 9.3).

Steam turbines are available in sizes from a few horsepower through millions of horsepower. It should be noted that the installation of a steam turbine typically involves the piping of steam (insulated), condensate (insulated), cooling water supply, and cooling water return. These services should be available nearby in sufficient quantity and quality to effect an economic installation. The overall efficiency of the steam turbine system may be degraded due to heat losses, leaks that are not promptly repaired, steam traps that are not monitored, and insulation that is not promptly replaced.

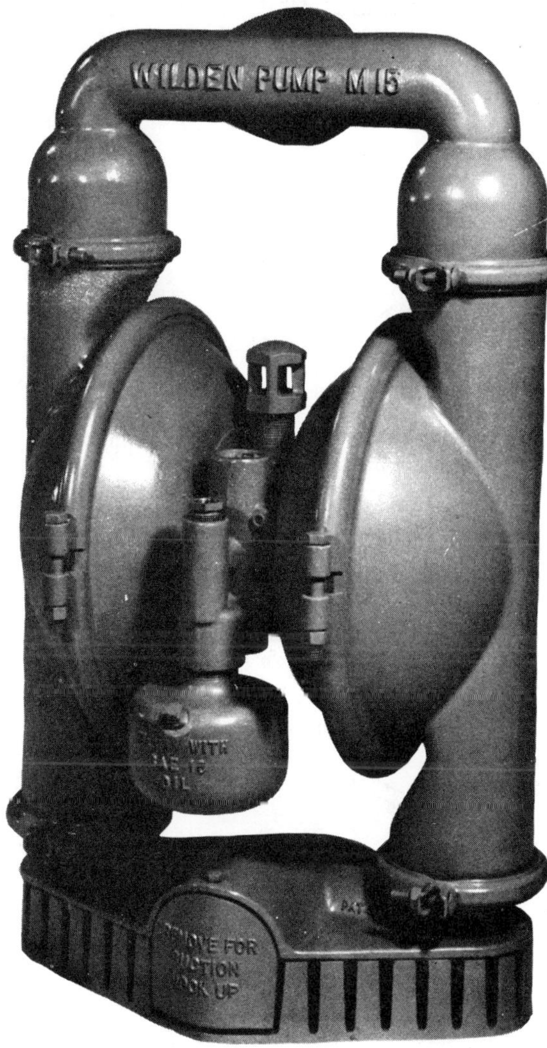

(a) Pneumatic Variable Speed Pump

(courtesy Wilden Pump and Engineering Co.)

Figure 9.2
Pneumatic Variable Seed Pump

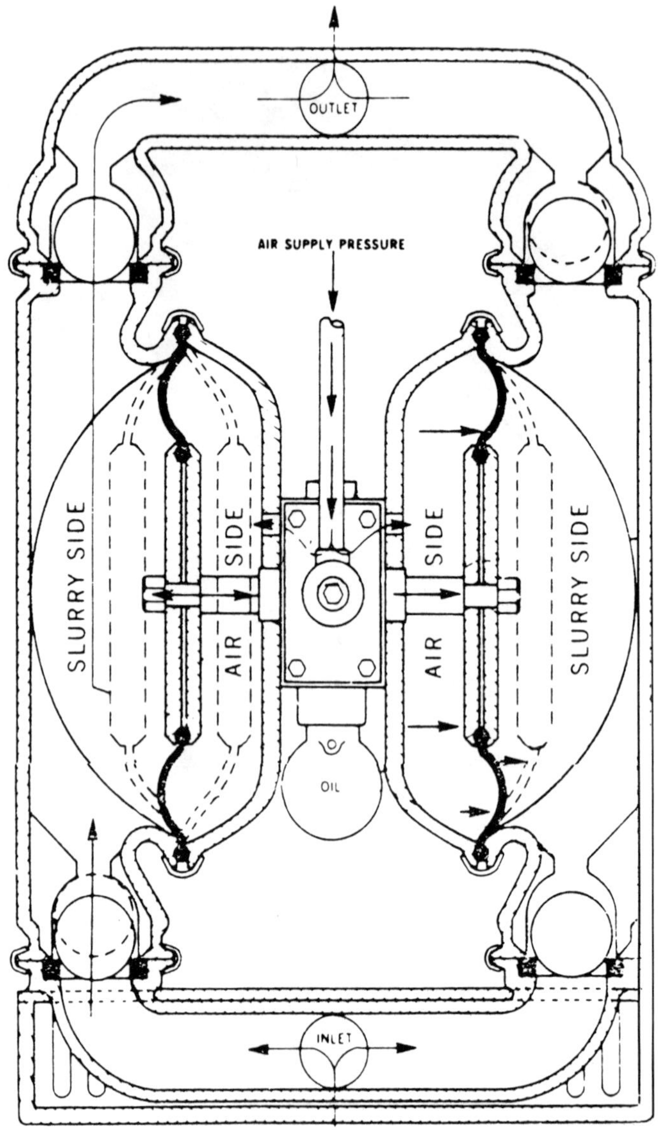

(b) Pneumatic Variable Speed Pump

(courtesy Wilden Pump and Engineering Co.)

Figure 9.2 (continued)

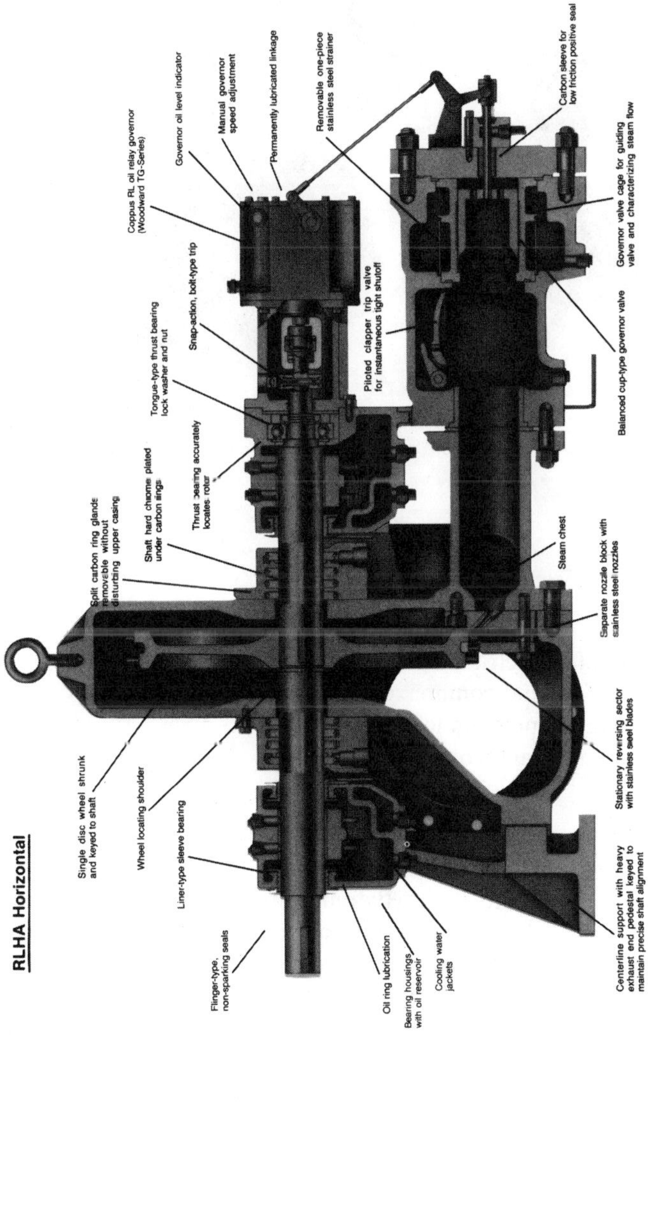

(a) Steam Turbine — Cutaway

(courtesy Coppus Engr. Corp.)

Figure 9.3
Steam Turbine

(b) Steam Turbine

(courtesy Coppus Engr. Corp.)

Figure 9.3 (continued)

Applications of steam turbines include the operation of:

Large generators for electric power generation.
Various pumps, fans and compressors located in the boiler house.
Pumps, fans and compressors located in hazardous areas.
Back–up pumps, fans and compressors.

Hydraulic Drives

Many varieties of hydraulic motors can be used to control the speed of large equipment (see Figure 9.4). Hydraulic drives can operate loads through tens of thousands of horsepower and satisfy high torque requirements. A motor is required to supply the pressure energy to the hydraulic fluid and may be supplied as part of a package or located remotely from the hydraulic motor to reduce exposure to the process environment. The hydraulic fluid is throttled to the drive with a valve, although variable capacity or variable speed drive technologies may be employed to achieve the same end. Hydraulic drives are usually compact but relatively costly to purchase and maintain.

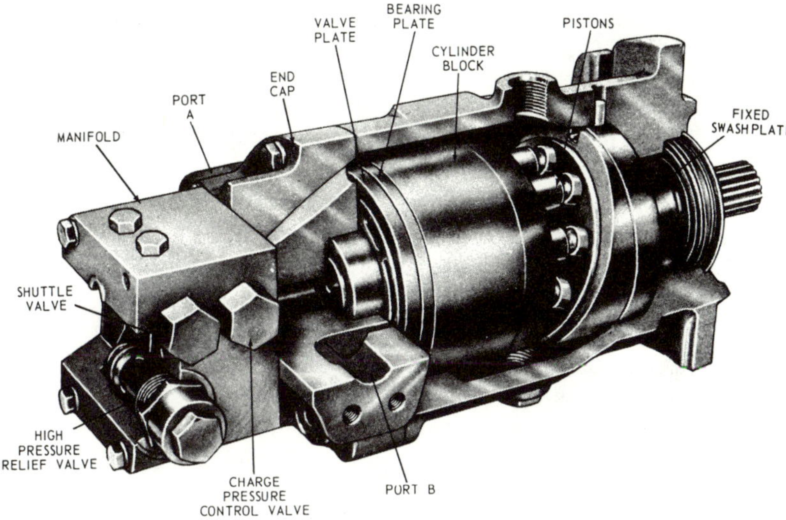

(a) Fixed Displacement Hydraulic Motor

(courtesy Sunstrand Sauer)

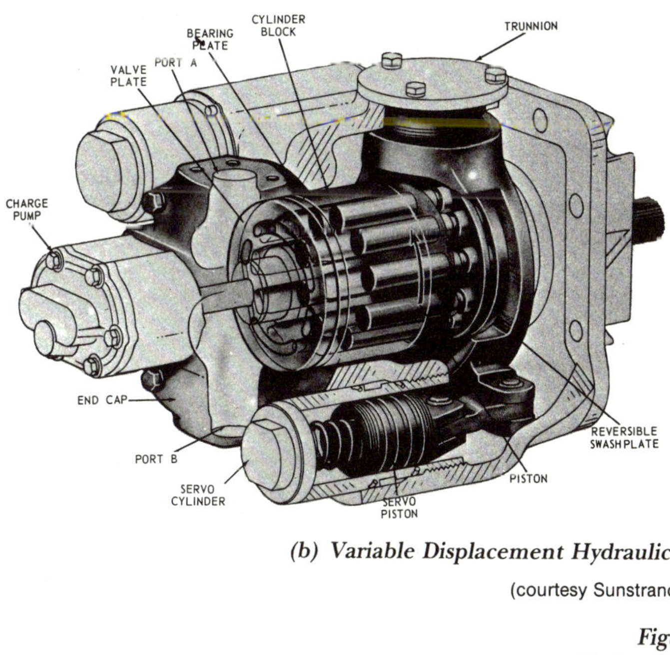

(b) Variable Displacement Hydraulic Pump

(courtesy Sunstrand Sauer)

Figure 9.4
Hydraulic Drive

FUELED PRIME MOVERS

Fueled prime movers encompass a number of technologies that transform the fuel to useful rotational energy. Speed is usually adjusted by controlling fuel flow. Fueled prime movers include internal combustion engines and gas turbines that may have single- or multiple-fuel capability.

Internal combustion engines are available from fractional horsepower through millions of horsepower and, depending on design, may be fired with a number of fuels including gasoline, diesel fuel, certain distillate oils, and natural gas.

Gas turbines are available from less than 10 horsepower through millions of horsepower in size and are typically fueled with natural gas, although some designs allow the use of certain distillate oils.

Applications of fueled prime movers include the operation of:

Larger pumps, fans and compressors.
Emergency power generation equipment.
Cogeneration equipment.

DRIVE COUPLINGS

A *drive coupling* located between a motor operating at constant speed and its load varies the speed of the load by allowing the output shaft of the drive

(a) Diesel Internal Combustion Engine and Generator

(courtesy Caterpillar)

Figure 9.5
Fueled Prime Movers

(b) Gas Turbine Engine for Automobiles

(courtesy Allison Gas Turbine Division, General Motors Corp.)

Figure 9.5
(continued)

to slip. The coupling, which can be thought of as a clutch, is typically implemented hydraulically or electrically (see Figure 9.6) and often applied to larger industrial motors.

Hydraulic couplings are devices that allow the amount of slippage between the input and output shafts to be adjusted hydraulically. This adjustment is implemented by varying the amount of hydraulic oil that is allowed to circulate in the coupling. Depending on size and manufacture, the oil circulation system can be extensive and the oil cooling requirements high, adding to the space and utility requirements of this technology. Hydraulic couplings are available from approximately 5 horsepower to millions of horsepower.

An electrical coupling, commonly known as an *eddy current clutch*, varies the speed of the load by allowing the amount of slippage between the input shaft (operating at constant speed) and the output shaft to be varied magnetically, generating eddy currents in the coupling. The speed of the load is dependent upon an electrical signal, which in turn alters the amount of magnetic coupling between the input and output shafts. Eddy current clutches are typically limited to speed adjustment of the load between 75 and 100 percent of motor speed, which is usually sufficient for centrifugal pumping applications. Air cooling of the coupling is usually sufficient to dissipate waste heat in sizes up to 3000 horsepower.

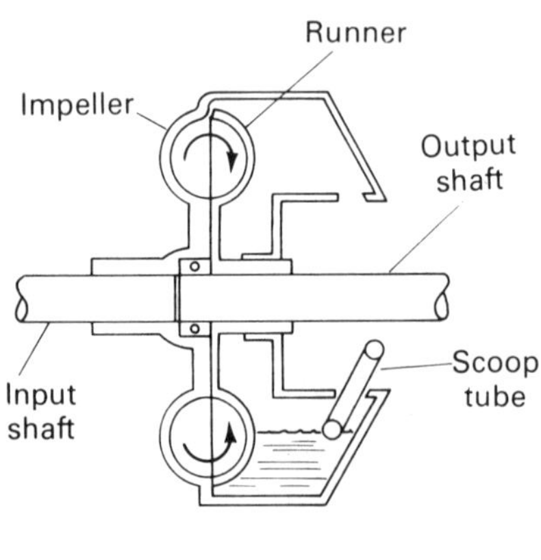

(a) Hydraulic Coupling Principle

(b) Hydraulic Drive Coupling

(courtesy American Davidson)

Figure 9.6
Drive Couplings

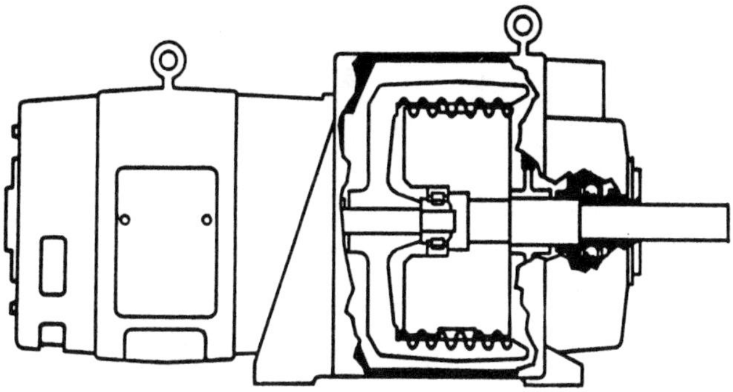

(c) Eddy Current Clutch

(courtesy Eaton — Electric Drives Div.)

(d) Eddy Current Clutch

(courtesy Eaton — Electric Drives Div.)

Figure 9.6 (continued)

DIRECT CURRENT MOTORS

Direct current motors have long been used to vary the speed of motors up to approximately 3000 horsepower by varying the voltage of the power source used to operate the motor.

In a typical dc variable speed drive, alternating current is rectified into direct current, the voltage of which can be varied. Motor speed is then variable as a function of the magnitude of the voltage of the dc power source. Due to the arcing and sparking nature of dc motor brushes, dc drives are usually not installed in hazardous locations.

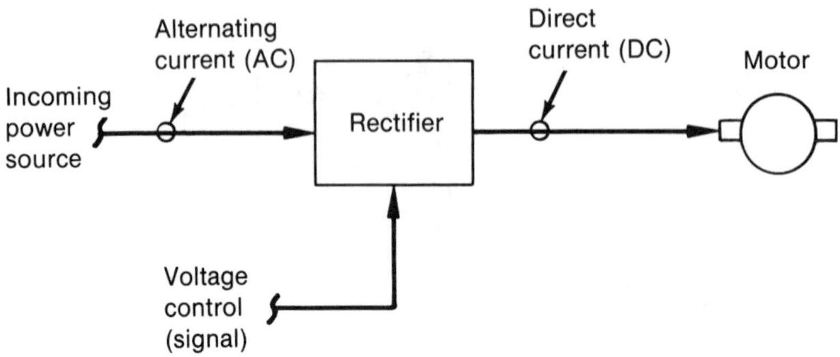

(a)

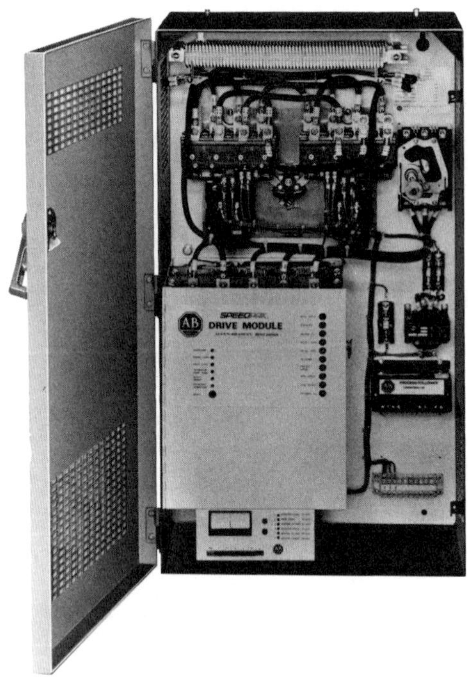

(b)

(courtesy Allen-Bradley)

Figure 9.7
DC Variable Speed Drive

WOUND ROTOR MOTORS

Motor speed can be varied by altering wiring that is normally considered to be internal to the motor. *Wound rotor* motor control is designed so that the resistance of the rotor winding is changed by externally connecting a variable resistor in series with the rotor winding, thereby altering motor speed (see Figure 9.8).

The power loss associated with the resistor can be significant at lower speeds and the waste heat from the resistor must be appropriately dissipated. Regenerative wound rotor drives utilize the energy that would

(courtesy Brown Boveri)

Figure 9.8
Wound Rotor Drive

normally be dissipated across the resistor to feed the power line. This is accomplished by (1) rectifying the current to dc, (2) inverting the current to an alternating current that matches the line frequency, and (3) transforming the voltage to match line voltage. Wound rotor variable speed drives are available from 1 to approximately 10,000 horsepower.

VARIABLE FREQUENCY DRIVES

The speed of a standard induction motor can be varied by adjusting the frequency and voltage of the power source to the motor. The incoming ac power source is rectified to effect a dc power source, which is subsequently inverted to provide ac power for the motor of the proper frequency and voltage.

Speed varies linearly with frequency where full motor speed occurs at line frequency. Some variable frequency drives can generate frequencies up to twice the line frequency. Voltage is generally maintained proportional to frequency to maintain the desired flux density within the motor. At low speeds, a voltage boost may be required in some designs to assist in achieving smooth motor rotation.

Variable frequency drives are available for motors through 4000 horsepower (see Figure 9.9).

ALTERNATE FINAL CONTROL ELEMENT SELECTION

Properly applied, alternate final control element technology will achieve some measure of success in controlling the operation of equipment in a plant environment. The degree of economic and operational success is in part dependent upon the alternate final control element selected.

The primary criteria used to select among alternate final control element technologies in a given application are the familiarity with the various final control elements, industry practice, technical requirements, reliability, and costs (installed, operating and maintenance). The effectiveness of the selected final control element for a given application should be evaluated by considering the performance and overall cost effectiveness of the installation.

Industry Practice

Historically, control valve technology was utilized for final control applications except those that exhibited unique technical requirements. Many of the alternate final control elements described above represent hardware that

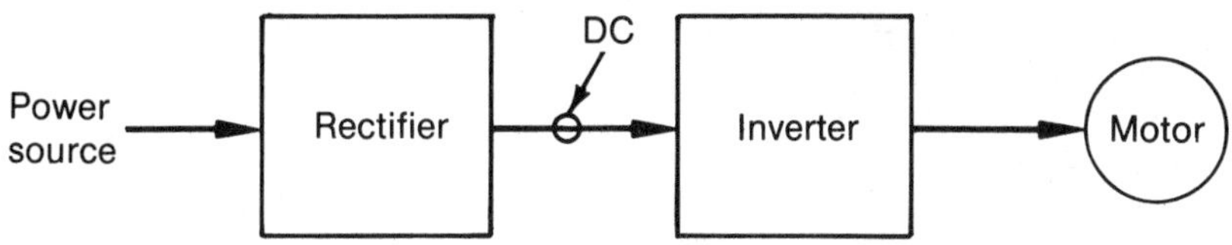

(a) Block diagram

(b) AC Variable Frequency Drive

(courtesy Robicon)

Figure 9.9
AC Variable Speed Drive

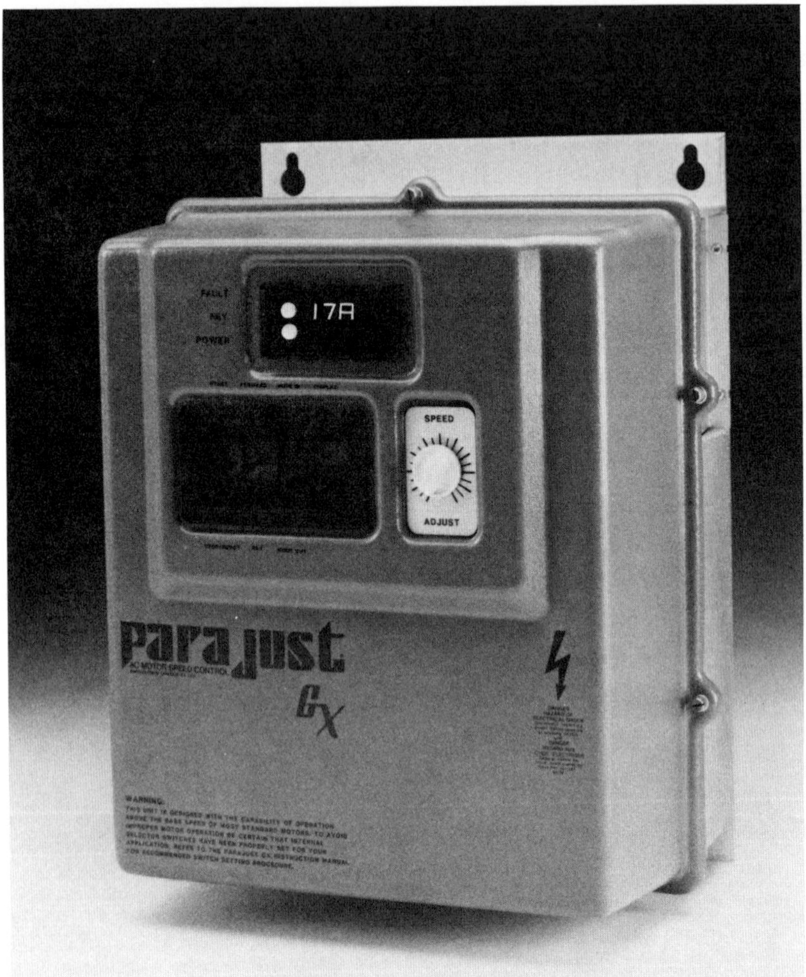

(c) Variable Speed Drive (10 hp)
(Microprocessor-based) Parajust GX

(courtesy Parametrics)

Figure 9.9
(continued)

has evolved over the years to meet these unique requirements. As much of this hardware was developed specifically for a given industry, it is not uncommon to encounter industries that have uniformly embraced technologies entirely different from those of other industries to achieve the same end.

Technical Requirements

The technical requirements of an application will, in most cases, eliminate many alternate final control elements. For example, a steam turbine would not be considered to power a load unless all utilities were readily available (as would be the case in the boiler house) or special design requirements existed (such as operation of the load during a power outage where no emergency power was available). A more practical example is the unsuitable application of dc drive technology to a dc motor located in a hazardous location due to the arcing and sparking nature of the motor brushes. On the other hand, the selection of alternate final control elements may be limited by the availability of space. In some applications, dynamic parameters, such as speed of response or immunity to abrupt load changes, may preclude the utilization of some technologies. Many examples can be given; however, each application must be considered on an individual basis with technologies eliminated or accepted accordingly.

Reliability

Reliability is one of the primary concerns of many users who apply any final control element. While it is difficult to fight success, technology that has been applied (seemingly) successfully and reliably for years should be considered acceptable only after the capabilities and trade-offs of currently available technology have been investigated and have been found to offer no substantial improvement in cost, performance, and/or reliability.

Costs

The costs associated with alternate final control elements include installation, operating, and maintenance costs. In the process of considering the various technologies, it is prudent to consider all of these costs as the rule rather than as the exception. In many cases, the operating and maintenance costs dwarf the installed cost of an alternate final control element.

Maintenance Requirements

Zero maintenance requirements is a goal that is rarely, if ever, achieved in practice. However, some technologies will require significant periodic and preventive maintenance, while others will not. For example, hydraulic systems must be periodically checked for leaks, oil level, oil quality, and the like, whereas electronic drives may require only annual inspection and cleaning.

The abilities of maintenance personnel may be limited in some locations; therefore, catastrophic failure of sophisticated equipment can often pose special problems. Should this be the case and all things being equal, it may appear to be prudent to install a simple device. While it may be more expedient in the short run to install a device that can be maintained by local personnel, the long run usually shows that failure to exploit technology results in a facility that is at an economic disadvantage. Continuing education and seminars, which may be generic or equipment specific, can provide maintenance personnel with the necessary skills to maintain sophisticated equipment, thereby overcoming the training problem in the long run.

Variable Frequency Drive Technology

10

Many alternate final control elements were considered necessary evils in the past when maintenance and operating costs may have been of little concern to the individual selecting alternate final control equipment. Present applications of this technology should be examined as an opportunity to improve the cost effectiveness of a process and reduce maintenance, which represents a radical shift in the light in which alternate final control elements are viewed.

While all of the alternate final control elements have their place, recent technical developments and cost reductions, primarily in the field of semiconductor technology, have resulted in an increased application of electronic variable frequency drive technology. The growth in the applicability of this technology has primarily been to transform applications previously considered neither practical nor feasible into viable projects. Various features inherent to its solid-state design have enabled variable frequency drive technology to displace other final control element technologies.

Variable frequency drive technology is examined below from a user's perspective to illustrate its capabilities and limitations as they relate to other alternate final control element technologies. It should be noted that variable frequency drive technology is not universally applicable, and that other technologies may be better suited for a given application.

SOLID-STATE DESIGN

Variable frequency drive technology employs solid-state electronic techniques to vary motor speed, thereby varying the operating speed of a piece of equipment. These drives have no moving parts and hence require minimal maintenance when compared to other non-electronic alternate final control elements.

VARIABLE FREQUENCY DRIVE TYPES

As the synthesis of a pure sine wave is prohibitively expensive, the output of the variable frequency drive approximates the sine wave at the desired frequency. The nature of the approximation depends on the semiconductor technology and the control strategy employed to generate the power output.

The design of these drives is such that semiconductor operation avoids the large amount of heat dissipation that occurs should the semiconductor be only partially conducting. Consequently, the semiconductors are switched on and off rapidly in an attempt to minimize heat generation. The net result is the generation of waveforms with short rise times and many harmonics, which are partially smoothed in the output stage of the drive.

While many smaller variable frequency drive applications control single-phase motors, most industrial applications utilize three-phase motors. In three-phase motors, the nature of the waveform fed to the motor windings is determined by the voltage difference across two of the phases, as opposed to a voltage difference between the output and ground as in the case of a single-phase motor.

As the semiconductors do not operate regulated, but rather on and off, the voltage waveform is generated by two semiconductors on each phase that connect that phase to either the + dc voltage or − dc voltage. The timing of the alternate connections to the dc system determine the shape of the waveform generated across the motor windings (see Figure 10.1).

The overall drive control strategy is to produce a waveform that is compatible with the motor, which in most cases tends to approximate the sine wave that the motor was designed to accept. One typically applied waveform is the 6-step waveform, which can be generated by alternately controlling the on and off states of the semiconductors (see Figure 10.2). Other waveforms can likewise be generated.

Variable Voltage Operation

Variable voltage inverters (VVI) utilize a rectifier that is capable of producing dc voltage and frequency that are proportional to speed and an

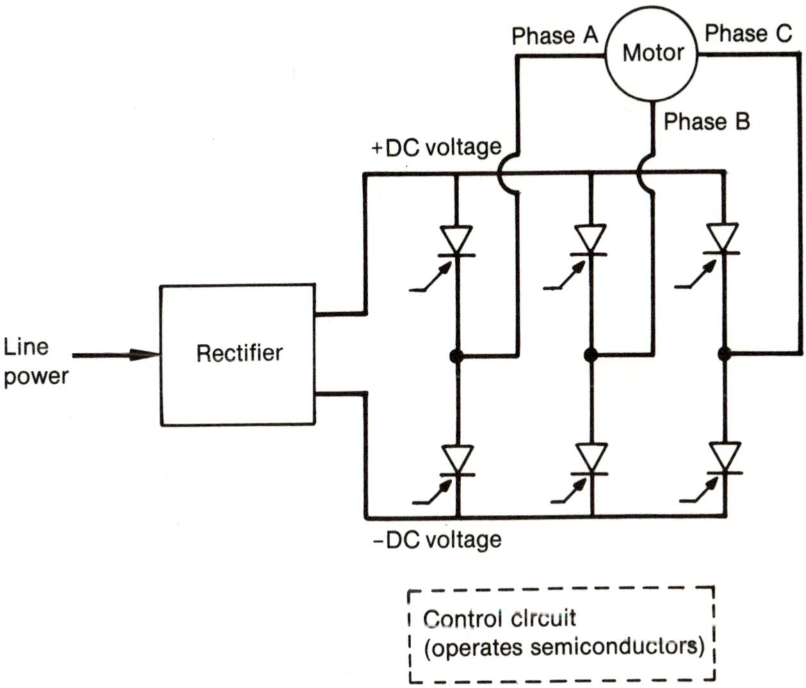

Figure 10.1
Simplified Variable Frequency Drive Circuit

output circuit that produces a 6-step waveform. The net result is an step-wise approximation of a sine wave voltage whose amplitude has been appropriately reduced to maintain the proper voltage to hertz ratio.

Current Source Operation

Current source inverters (CSI) utilize a rectifier capable of producing a dc voltage that is proportional to speed and an output circuit that controls the current to the 6-step waveform output. Current control allows the drive to have complete control of motor torque at the expense of more complex control.

Pulse Width Modulation

Pulse width modulation (PWM) drives utilize a rectifier that produces a constant dc voltage independent of speed. The output circuit controls both

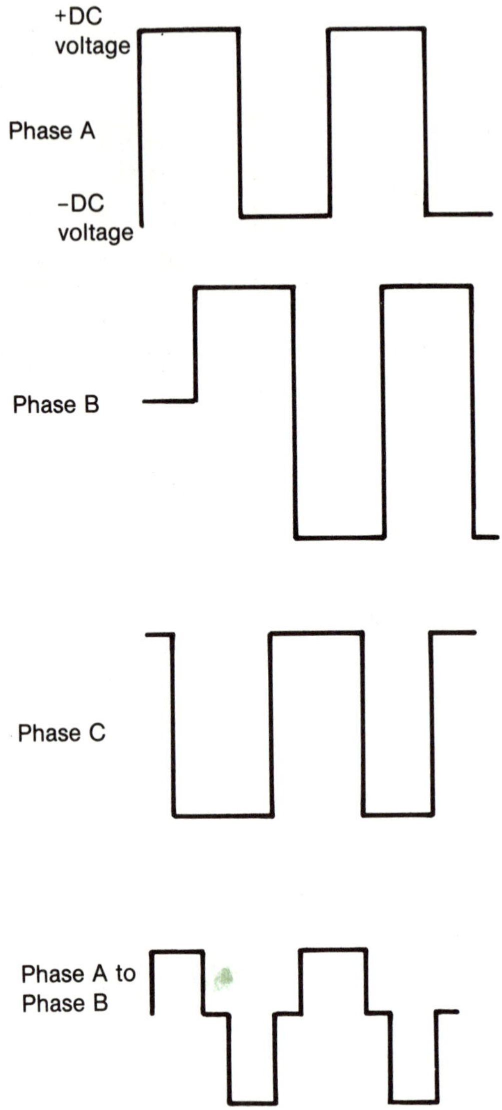

Figure 10.2
Generation of a 6-Step Waveform

the voltage and the frequency by varying the width of the output pulses (see Figure 10.3). This type of variable frequency drive requires the most complex switching control and regulation.

HORSEPOWER AND TORQUE

In the constant torque mode of operation (below line frequency), the available *horsepower*, or work that can be done, is a linear function of motor speed. In this speed range, the *torque*, or rotational energy that can be produced at the motor shaft, is constant and corresponds to the full rated torque of the motor at line speed (see Figure 10.4). Control of torque is accomplished by maintaining a constant voltage-to-frequency ratio, except at very low speeds where the motor may require an increased voltage to operate properly.

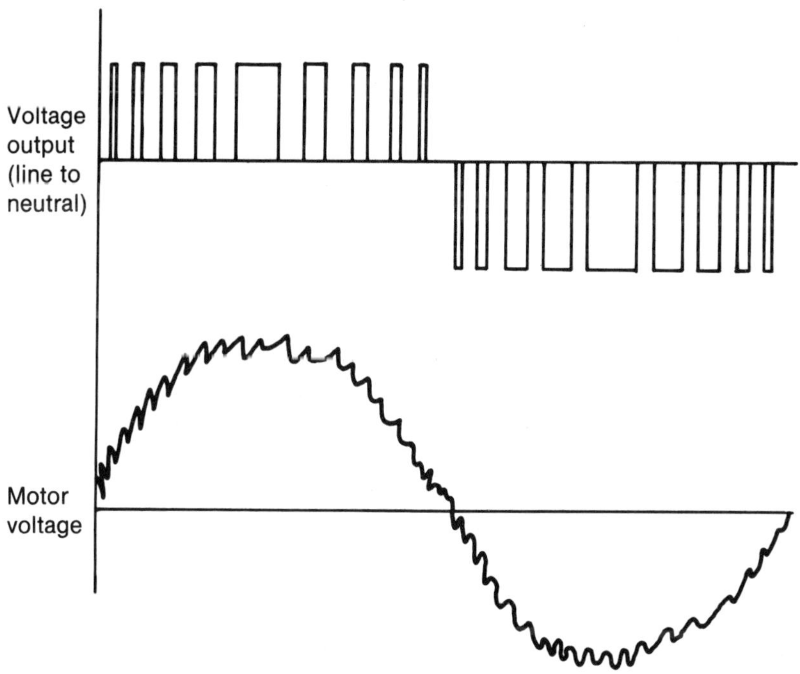

Figure 10.3
Typical Pulse Width Modulation Waveforms

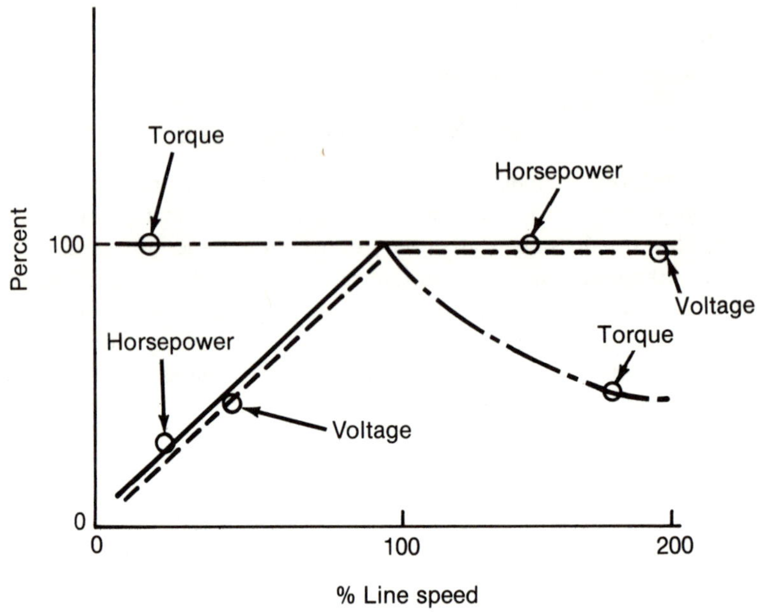

Figure 10.4
Relationship between Horsepower and Torque in Variable Frequency Drive Operation

In the constant horsepower mode of operation (above line speed), the voltage is at its maximum, and the horsepower is constant and limited by the motor to the rated horsepower at line speed. The torque that the motor is capable of producing is reduced accordingly. In this speed range, steps should be taken to verify that the motor and equipment will operate satisfactorily above line speed. Operation of any equipment over 4000 revolutions per minute (rpm) should be thoroughly investigated.

TYPES OF LOADS

Centrifugal loads and *constant torque loads* represent the great majority of loads that are operated by variable speed drives. The operating characteristics of each type of load are different. Due to torque requirements, centrifugal loads will operate properly with constant torque drives, but constant torque loads will trip a centrifugal drive under certain conditions.

Constant Torque Loads

Positive displacement compressors and positive displacement pumps are typical of loads that require constant motor torque regardless of motor speed. Extreme care should be exercised when designing a system to operate a constant torque load above line speed, as the torque that the drive/motor system will produce is reduced in the constant horsepower mode. The required motor horsepower is a linear function of speed both above and below line speed (see Figure 10.5).

Below line speed in the constant torque mode, the motor tends to remain fully loaded and draw full load current, regardless of speed.

Centrifugal Loads

Fans, centrifugal pumps, and the like typically follow the Affinity Laws where the horsepower requirement is proportional to the cube of the speed,

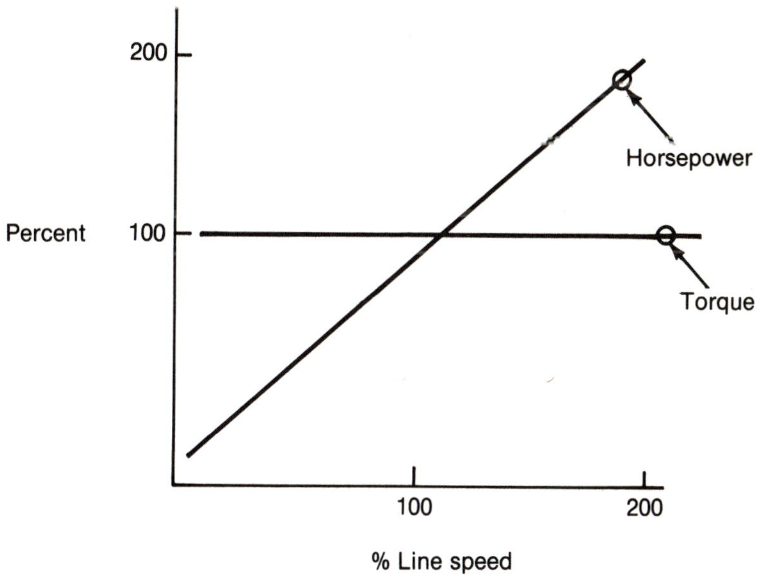

Figure 10.5
Horsepower and Torque Requirements of a Constant Torque Load

that is, doubling speed will increase horsepower requirements 8 times. Conversely, reducing motor speed from 100 to 90 percent of line speed will result in a reduction of the required horsepower of $1-(0.9)^3$, or 27 percent.

As power requirements are proportional to the cube of the motor speed while motor capacity is proportional to motor speed, below line speed the motor operates partially loaded at less than full load current, efficiency and power factor (see Figure 10.6).

Extreme care should be exercised when designing a system to operate a centrifugal load above line speed, the motor must be sized and its speed controlled so that motor capacity is always greater than the load.

POWER FACTOR CORRECTION

In general, power factor correction is a desirable feature that can affect the electric bill. Most variable frequency drives contain power factor correction and operate at a power factor of approximately 94 to 96 percent at full load,

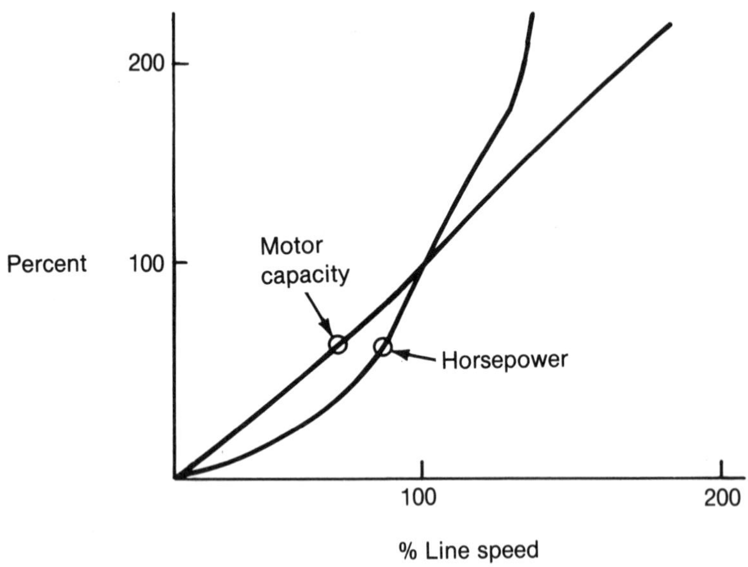

Figure 10.6
Horsepower Requirements of a Centrifugal Load

but may be significantly lower at lower loads. Most smaller drives are designed to operate in this region under all load conditions. Depending on size and manufacture, some drives may contain no power factor correction equipment at all, which, when applied to constant torque and many centrifugal loads, can increase the electric bill.

The power factor of pulse width modulated drives is usually optimized to be near unity regardless of motor speed, due to the design of its constant voltage source. The power factor of variable voltage and current source drives decreases proportionally as speed is decreased, due to the phase angle of the power output of the drive. Therefore, kVA demand of a fully loaded motor (regardless of motor speed) will be that of a motor that is operating fully loaded at full speed. Some voltage source and current source drives are designed to operate at a high power factor independent of load.

Power factor correction equipment can be economically advantageous by reducing the current required to operate the drive at reduced speeds, thereby reducing transformer loading. This reduces the transformer and cable heat losses, as well as kVA demand charges.

The economic benefits of power factor correction equipment should be examined to determine the viability of additional expenditures for this purpose. It should be noted that when traditional power factor equipment is used (capacitors and inductors), the power factor will vary somewhat as a function of speed.

EFFICIENCY

The efficiency of a variable frequency drive is the ratio of the electric power delivered to the motor to the input power consumed by the drive. The measurement of these parameters is relatively difficult due to the non-sinusoidal nature of the input and output current waveforms. However, efficiency can be thought of in terms of the heat losses associated with the power semiconductors and the relatively constant heat loss associated with the electronic circuitry. These losses can represent approximately 2 to 6 percent or more of full load power in typical variable frequency drives. The majority of the heat losses associated with medium and large drives are due to power semiconductor heat effects and vary as a function of motor current.

More important to the user is the efficiency with which the variable frequency drive utilizes billed electrical consumption to produce brake horsepower and, ultimately, hydraulic energy. First, the power factor correction feature of most drives will maintain a near unity power factor regardless of the actual motor power factor, effectively reducing kVA demand. Second, in many applications proper instrumentation and control can keep the motor fully loaded, thereby maintaining optimum motor efficiency at reduced speeds.

Detracting from overall efficiency are the effects of synthesized non-sinusoidal waveforms on motor operation. The relatively short rise time of these waveforms creates harmonics, which result in some heating effects that adversely affect motor efficiency.

It should be clearly noted that the ultimate measure of efficiency to the end user is system efficiency, that is, the ratio of hydraulic energy to billed electrical energy, which focuses on utilization of electricity in a cost-effective manner. For example, at full speed and full load, there is virtually no difference in the efficiency of the various types of variable frequency drives; however, the current source inverter tends to maintain efficiency as speed is reduced, the economic effects of which may be partially offset by the increased power factor offered by a pulse width modulating drive that maintains a near unity power factor regardless of speed. In short, it should be noted that no one type of drive is universally applicable.

An estimate of typical losses associated with a pulse width modulating drive is approximately 1 percent of full load plus 2 percent of actual load. The efficiency of the drive at a given load can be estimated as

$$\text{Efficiency (percent)} = 100 - (2 + 1 \times \text{Load/Full load})$$

MOTOR REQUIREMENTS

One of the primary advantages of variable frequency drive technology is that it can be applied to induction motors that are commonly used in hazardous and nonhazardous locations throughout industry. Virtually any induction motor can be driven with a variable frequency drive; however, there are some special considerations that must be taken into account when applying one.

It should be recognized that motors generate substantial amounts of heat during operation. This heat must be dissipated to maintain cool motor operation and motor longevity. Motor operation at reduced speeds results in a reduction of fan cooling effectiveness, which can cause the motor to heat excessively and can bring about premature motor failure. Harmonics, due to the short rise times of the synthesized waveforms, add to motor heat dissipation requirements.

The most operationally flexible solution to the heat dissipation problem is to utilize a motor with a 1.15 service factor or higher insulation class that is designed to operate under the above conditions.

When the motor does not have the required service factor, the motor should not be operated at full rated torque regardless of speed. Depending on the nature of the load, this is accomplished by limiting the maximum speed at which the motor will operate. In other applications, appropriate adjustment of the load or control algorithm may alleviate the problem. The net result is a derating of the drive/motor system (see Figure 10.7).

As fan-cooled motors can get excessively hot when operated at less than 10 to 30 percent of line speed, depending on the characteristics of the load, it may be appropriate to limit the minimum speed of the motor. In most applications, the motor should not be allowed to operate fully loaded at less than 10 percent of motor speed.

SPEED ADJUSTMENT

Adjustment of motor speed may be accomplished in a number of ways, depending on the application. Most variable speed drives offer manual as

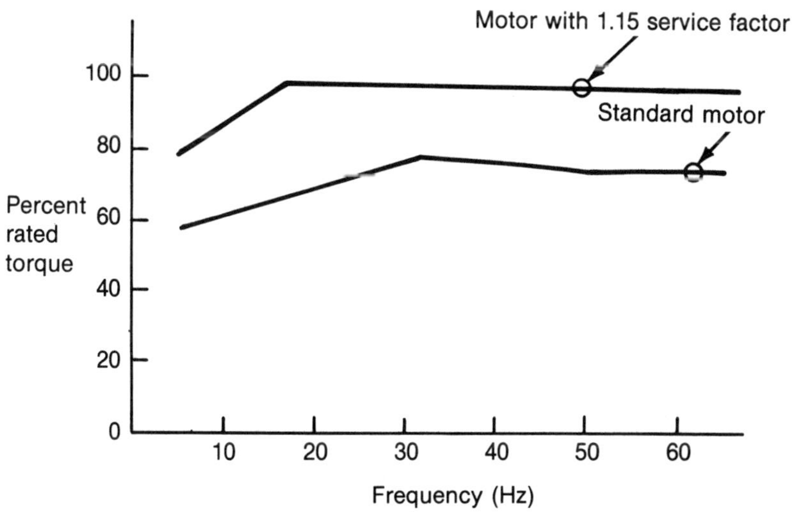

Figure 10.7
Typical Effect of Reduced Fan Cooling on Motor Torque

well as remote speed adjustment, utilizing a 4–20 milliamp signal from the instrumentation system. Variations on the above, such as the ability to accept a pneumatic signal or digital interfaces, may be available from some manufacturers.

Most drives incorporate minimum and maximum speed adjustments that the instrument signal cannot override. This feature is particularly useful where operation of the equipment outside these parameters can damage equipment or cause a potentially unsafe condition to exist. These adjustments can also be used to modify the control range to attain better resolution of speed, such as where 4 to 20 milliamps may be adjusted to correspond to 60 to 100 percent of speed instead of the usual 0 to 100 percent. Note that not all drives allow such large zero and span adjustments.

START/STOP/REVERSE CIRCUITS

As a minimum, variable frequency drives should be powered from a disconnect switch, which allows maintenance personnel to completely and safely disconnect the drive from the power line for repair without de-energizing other circuits as well. Most variable frequency drives contain start/stop circuitry that eliminates the need for a combination starter. Most manufacturers will incorporate disconnects and/or starters into the design of the drive when specified.

Variable frequency drives inherently produce a soft motor start. While a motor connected across the line will accelerate rapidly to full speed and draw up to 6 times its full load current, variable frequency drives accelerate the motor at a controlled rate that is adjustable and limited by the current capability of the drive. Full motor torque is available during motor acceleration.

Variable frequency drives decelerate the motor at a controlled rate by putting an electrical load on the motor, thereby dynamically braking it. Some drives are designed to recuperate the rotational energy of the motor, using regenerative braking circuits that generate electricity during braking and put power back onto the line. These drives are more complex and may be economical when the motor is constantly accelerated and decelerated.

Alteration of the firing sequence of the power semiconductors allows many variable speed drives to operate a motor in the forward and reverse directions. Reversing logic circuits are usually incorporated into the drive, which ramps the motor to a halt before accelerating the motor to the desired speed, in the opposite direction.

MOTOR/DRIVE PROTECTION

Variable frequency drives are electronic devices that supply current to the motor, which is limited by the current rating of the output semiconductors. Under overload conditions, the variable frequency drive will trip itself to protect the drive and the motor. Depending on manufacturer, variable frequency drives may also trip as a result of other abnormal conditions, such as ground fault, over-voltage, and the like. Dedicated lights for each fault are available from some manufacturers, while others utilize diagnostic cards to display a code that identifies the fault. In any case, proper diagnostic indication can greatly enhance the ability of maintenance personnel to identify and correct a problem.

LOCATION

Variable frequency drives may be located at the motor location or remotely, when required. In industrial applications, drives are typically located in a motor control center remote to the process so they will not be exposed to the process environment, which may be hazardous and/or corrosive. In many commercial applications, the variable frequency drive is located in the proximity of the motor.

MULTIMOTOR OPERATION

Properly sized, some variable frequency drives are capable of operating more than one motor connected in parallel, while multiple drives may be capable of synchronizing the motor speeds of various motors. Motor speeds can also be controlled to be a ratio of one another within certain accuracy limits. It is well advised to consult with a manufacturer when considering multimotor variable frequency drive applications.

Control Valves vs. Variable Speed Drives

11

While control valves are the most widely applied final control elements in industry and, in all probability, will continue to dominate final control element applications for many years to come, variable speed drive technology also has a legitimate and expanding role in industry. Historically, alternate final control elements have been viewed as necessary evils. However, due to technological advances, many applications now represent technical opportunity and the possibility of significant economic reward.

As neither the control valve nor variable speed drive technologies are universally applicable, a comparison between them should prove helpful in determining the advantages and disadvantages of each technology in a given application. Depending on the application, the concepts presented here may be divided into (1) quantifiable items that can be measured and (2) so-called "motherhood and apple pie" items that are known to provide concrete results but cannot be measured due to a lack of pertinent quantitative information.

Initial thought on an application should focus on the conceptual advantages and disadvantages of a particular technology. The following comparison focuses on the conceptual differences between the technologies that may be applicable in making the initial decision to consider variable speed drive technology or to use a control valve in a given application.

CONTROL PHILOSOPHY

Perhaps the most fundamental difference between the control valve and variable speed drive technologies is the philosophical difference between the type of control each is capable of offering and the ramifications thereof. It should be clearly understood that a control valve is a device that dissipates hydraulic energy, while a variable speed drive is a device that can regulate the amount of hydraulic energy generated.

In an analogy, the application of a control valve can be thought of as the case where a motor vehicle's engine is operated continuously at full speed while the brake is manipulated to control the speed of the car. Conversely, variable speed drive technology can be thought of as operation of the vehicle when the brake is totally released and engine speed is used to control vehicle speed.

EFFICIENCY

One fundamental divergence between control valve and variable speed drive technologies is that the former entails the generation of hydraulic energy only to dissipate excess energy not required by the load, while the latter generates only the required amount of hydraulic energy necessary to match the load. The net result is that in an application where either technology can be applied the implementation of variable speed drive technology tends to result in a system whose energy input better matches load, making the overall system more efficient.

POWER FACTOR

While the power factor of a motor operating at full speed in a control valve application can be over 90 percent in larger motors operating under full load conditions, it can drop off drastically as load is reduced. Most electronic variable speed drive designs incorporate power factor correction circuits, which maintain near unity power factor regardless of load. In most applications, the improved power factor reduces kVA demand charges and transformer loading.

OPERATING COSTS

In typical applications, variable speed drive technology offers the user increased efficiency over a control valve. The net result is that variable speed drives can reduce the electric bill and will often yield electrical savings sufficient to justify most new applications as well as some retrofit applications.

LOCATION

In the great majority of applications, control valves must be located in the field and are subjected to the process environment, which may be hazardous, corrosive, hot, cold, wet, or snowy. Electronic variable speed drives may be located either in a motor control center, where the drive will be protected from the process environment, or in the field, if required, making the location of a variable speed drives more flexible in this regard.

INSTALLATION

The installation of a typical pneumatic control valve in a continuous operation involves giving consideration to the installation of bypass piping. Air supply piping and signal wiring from the instrumentation system compete for space in the process area. The installation of a variable speed drive requires sufficient space and access in the motor control center or in the field, if so located. Some electronic variable speed drives may require that the cables to the motor be routed separately from the incoming power cables. In many cases, variable speed drive installation is more straightforward, although less practiced, than the installation of a control valve.

Retrofit electronic variable speed drive applications can be installed in a straightforward manner when sufficient space is available. In most cases, the existing cables connecting the motor to the starter can be reconnected to the drive and reused. It should be noted that there are many applications that may require extensive electrical and mechanical work, such as where space for the drive is not available or major electrical rework must be done due to the existing cable routing.

While electronic variable speed drives are typically located in a protected environment, the wetted parts of a control valve are exposed to the process fluid, which may be corrosive and/or abrasive and subject to physical wear and damage. While appropriate materials of construction can alleviate these problems, the valve must eventually be rebuilt due to age. The variable speed drive effectively eliminates one area of potential exposure of the fluid to people and the environment.

It should be noted that in some applications piping considerations must be made to enable the variable frequency drive and controls to function properly in the system.

DESIGN

While a typical control valve installation will involve a review of process piping arrangements, instrument air distribution, and conduit layouts, an electronic variable speed drive requires sufficient access and space in the motor control center. In this case, development of the layout is similar in difficulty to that of a control valve; however, the design work associated with other variable speed drive technologies may be substantially more involved.

SPECIFICATION

However mundane a control valve application may be, the valve specification represents the product of many technical and economic decisions and trade-offs. In a typical control valve application, substantial amounts of process data and compatibility information must be accumulated in order to size and select a control valve. As a minimum, control valve selection entails specification of valve type, valve trim and its inherent characteristic, materials of construction for all wetted parts, accessories, and the like.

Specification of electronic variable speed drives is usually simpler than that of a control valve because it is primarily dependent on the motor and type of load, information about which is more readily available than is process data and compatibility information. The specification of other variable speed drive technologies can be considerably more complex than that of specifying a control valve. The immunity of variable speed drive technology to the problem of material compatibility represents a distinct technical advantage over control valve technology.

OPERATION

While control valves are capable of completely closing, stopping flow in both directions (within the valve leakage specifications), variable speed drive equipment can do nothing of the sort. For example, when the head supplied by a pump (fan) is insufficient to overcome the head in the piping system, the flow will be reversed through the pump (fan). Reverse flow (which may affect flowmeter performance) can be minimized by installing a check valve at the discharge; however, stopping the motor does not create zero flow and positive shutoff, as does a control valve.

In short, where positive shutoff is required, control valve technology should be considered, but a variable speed drive with an appropriately controlled on-off valve may also be applicable.

Control valve applications typically include a bypass arrangement that allows the removal of the control valve for service while operating the bypass valve manually. It should be noted that experience has shown that, in many cases, bypass valves and the control valve tend to fail to operate properly after approximately the same length of service, often necessitating a process shutdown for repairs when the problem is intolerable.

A similar arrangement can be designed to remove a variable speed drive from service where spare equipment is operated at line speed with its discharge throttled with a manual valve. Some electronic variable speed drives can be purchased with an across the line starter that enables the motor to operate while the drive is out of service.

In control valve applications, rotating equipment operates at full speed with full equipment capacity available at all times. On the other hand, a control system that incorporates a variable speed drive will control motor speed to match load. As a result, under abnormal or upset conditions the equipment may be momentarily incapable of overcoming the head of the piping system, resulting in reverse flow through the equipment (with a possible motor reversal) and potential equipment damage, process upset, and/or environmental emissions. Assuming that all equipment is functioning properly, eliminating the cause of the upset condition or modifying the control strategy can usually alleviate this problem.

ABILITY TO CONTROL

Inherent sluggishness can cause instability of the control loop, due to the time required for the control valve to react, and the dead band associated with valve positioning. The I/P converter introduces a time delay between the controller output and the pneumatic signal to the actuator. Friction caused by the actuator and packing causes the control valve to ignore small controller output changes encountered when the process variable is near its set point, or when the process reacts slowly. The resultant hysteresis causes the process variable to wander near the set point or be unstable. Control valve positioners are typically used to minimize these problems, although some delays remain.

Some alternate final control elements, notably electronic variable speed drives, control motor speed with negligible converter dead time and virtually infinite resolution. This eliminates most of the hysteresis problems associated with control valves, thereby improving control of the process variable.

SAFETY

Control valve leakage may be a minor inconvenience in many common applications, such as in water service, but leaks in more demanding service may present a potential hazard to life and property. Potential leaks are not limited to the control valve only; bypass valves are also potential sources of leaks. One advantage of variable speed drive technology is that it typically adds no equipment to the piping system, thereby minimizing potential sources of leaks. However, variable speed drive technology may be detrimental to plant safety in those applications where a drive failure introduces special safety considerations.

INSTALLED COST

While the purchase price of a small (under 10 horsepower) variable speed drive for a new application is significantly higher than that of a typical 1 or 2 inch (25 or 50 millimeter) carbon steel control valve with accessories, its installed cost is generally less than or equal to that of the control valve after the equipment and accessories, bypass valves, air piping, conduit, cable, starter, breaker, labor, and the like are taken into account. Retrofit applications utilizing variable speed drive technology are typically more costly than replacing a control valve, as the electrical and pneumatic utilities are already installed.

The installed cost of larger drives can greatly exceed that of a control valve; however, the potential operating cost savings are usually large as well. In some variable speed drive applications, the size of the equipment and motor can be smaller than that of a control valve application due to the reduced hydraulic energy requirements.

MAINTENANCE

At a minimum, the maintenance costs associated with control valves include fixing leaks and periodically rebuilding the valve, either in the field or after removing the control valve from the line and transporting it to the shop. Valve accessories, such as current-to-pneumatic converters and positioners, should be periodically calibrated and checked for proper operation. The amount of maintenance required for a given number of valves is difficult to assess accurately. One rule of thumb states that the maintenance expenditures associated with a valve in 5 years of operation will equal its purchase price.

The electronic troubleshooting skills necessary to repair electronic variable speed drives are very different from the mechanical skills necessary for control valve repair. On the other hand, such a drive located in an area protected from the process environment should exhibit a relatively good maintenance record, although this is difficult to quantify. The conditions under which maintenance and troubleshooting are performed on electronic variable speed drives are generally better than those of a control valve, with the added benefit that they do not require the piping to be opened, thus minimizing the exposure of personnel to the fluid in the pipe. Maintenance on variable speed drives utilizing other technologies can be extensive.

The application of electronic variable speed drives will tend to reduce the amount of equipment maintenance, as reducing the speed and maintaining a relatively constant load will reduce the wear and tear on the equipment. This reduction in maintenance is difficult to quantify, since it varies with the application.

SPARE PARTS

Spare parts requirements for control valves can entail many specialized items due to the different valve types, sizes, and materials of construction, even with standardization of manufacturer throughout a plant. As electronic variable speed drive technology need not address materials problems, the number of spare parts may be fewer than those for control valves and will consist of more universal parts. The initial spare parts for these drives may be more expensive to purchase than those for a typical control valve, but a faulty part can often be returned to the manufacturer for repair or be repaired at the plant. Spare parts for other technologies can also be extensive.

Pumps operated with variable speed drives can use a standard size impeller, which eliminates the need to cut the impeller should an impeller change be necessary. A larger than the desired size impeller can be used by appropriately limiting the maximum speed of the drive to avoid overloading the motor. In some cases, fewer impellers need be stocked as spares.

SUMMARY

The following is a summary of the comparative performances of a typical application where a control valve or an electronic variable speed drive would be applicable. This selection of variable speed drive technology represents a widely applicable technology with a good price/performance ratio.

Item		Control Valve	AC Variable Frequency Drive
Equipment efficiency		—	Better
Motor efficiency		—	Better
Power factor		—	Better
Operating costs		—	Better
Flexibility of location		—	Better
Exposure to process		—	Better
Ease of installation		—	—
Specification		—	Better
Shutoff capability		Better	—
Ability to control		—	Better
Potential for leaks		—	Better
Installed cost:	small	—	Better
	large	Better	—
Maintenance:	valve/drive	—	Better
	equipment	—	Better
	expertise	Better	—
	spare parts	—	—

From the above table, it can be seen that neither technology is superior in all respects, but some subtle observations can be made:

(1) AC variable frequency drive operation is typically more efficient than control valve operation; however, the economic impact may or may not justify its installation.
(2) The application of variable frequency drive technology can reduce the number of pieces of equipment that are exposed to the process fluid.
(3) The installed cost of a variable frequency drive in applications of small motors is usually less than that of a control valve installation.
(4) More expertise is usually required to troubleshoot variable frequency drives than control valves.

(5) Whereas control valves typically provide shutoff capability, variable frequency drive applications may require piping changes to prevent backflow and provide tight shut–off capability.
(6) AC variable frequency drives provide superior control characteristics due to good resolution and negligible dead time.

Other variable speed drive technologies can be analyzed in a similar manner.

12 Typical Variable Speed Drive Applications

The initial impetus for a viable variable speed drive application should be based upon an in-depth understanding of the process to be controlled, as well as the advantages and limitations of variable speed drive technology as it relates to the application at hand. Conceptual design of a variable speed drive system is a multi-discipline process, initiated by the desire for economic advantage and/or superior operation. These basic goals should be addressed during the conceptual phase of applying variable speed drive technology.

Identifying conceptually viable variable speed drive applications is more an art than a science, but two primary concepts can be applied to better organize the process of identification.

First, the practice of producing hydraulic energy only to dissipate a portion of it unproductively is energy wasteful, especially as compared to producing only the required amount of hydraulic energy to match load. For example, a variable speed drive that generates the desired amount of hydraulic energy by varying motor speed is more energy efficient than is the same system operating at full speed while its discharge is throttled to deliver a similar amount of hydraulic energy.

Second, in some situations, the application of a variable speed drive may be the key to process changes and/or superior control that result in significant economic benefits. For example, eliminating oscillations in a control loop may allow the set point to be set closer to constraints, perhaps resulting in more economical operation.

To develop techniques with which to identify potential applications in a conceptual sense, the left side of the figures in this chapter illustrate simplified systems typical of those found in industry, while the right side shows the equivalent loop with a variable speed drive applied.

In a new facility, these control schemes usually appear in the design documents on the Process and Instrumentation Drawing (P&ID). While these loops may appear on the design documents of an existing plant, field modifications can and do occur, such that field verification of existing systems is imperative, especially the manual adjustment of the drain and shutoff valves. As a result, such manual adjustments, often a result of field modifications, must be found on field trips rather than by examination of design drawings.

GENERAL GUIDELINES

The primary techniques for identifying potential applications is to identify those systems that exhibit low efficiency in the generation of useful energy, or those where the quantity of energy provided is greater than the process requires. Some pointers for identification of potential applications are the following:

(1) Compare the actual operating capacity to the system capacity; systems operating at low loads are usually less efficient than at higher loads.
(2) Determine whether the suction or discharge piping is throttled, which is indicative of energy losses and reduced system efficiency.
(3) Understand the purpose of the rotating equipment in the process in order to determine whether any of the attributes of the technology will adversely affect the process if not properly addressed.
(4) Determine whether the equipment can be operated at reduced speed without damage.
(5) Determine whether a reduction in equipment operating speed improves process economics, either directly or indirectly.

Opportunities for improving energy efficiency usually exist when the system is operated for the majority of time at less than full system capacity and/or when the discharge piping is throttled or bypassed. The benefits of superior control and the process improvements associated with a variable speed drive must be examined on an individual basis.

PUMPING APPLICATIONS

Pumping applications represent a significant opportunity for applying variable speed drive technology in new as well as retrofit applications.

In a typical level control loop, shown in Figure 12.1a, the level in a vessel is controlled by throttling a control valve at the pump discharge, which is typical of a distillation column bottoms level control loop. In many applications, the flow through the pump will be between 25 and 75 percent of capacity the majority of the time. While this loop is conceptually straightforward, a significant percentage of the hydraulic energy generated by the pump is dissipated across the control valve to regulate the pressure downstream of the control valve to produce the desired flow. The pump must be sized to accommodate the pressure drop associated with the control valve at maximum flow.

Application of a variable speed drive as the final control element in the loop will control the pump so as to generate only the hydraulic energy required to discharge the desired amount of liquid. This approach reduces electric costs while reducing pump maintenance requirements. The net result is a system that reduces operating and maintenance costs by eliminating the need for a control valve, bypass piping, and the associated energy losses.

Figure 12.1b illustrates a control loop in which the flow discharged from a vessel is controlled, which is typical of a raw material feed from a storage

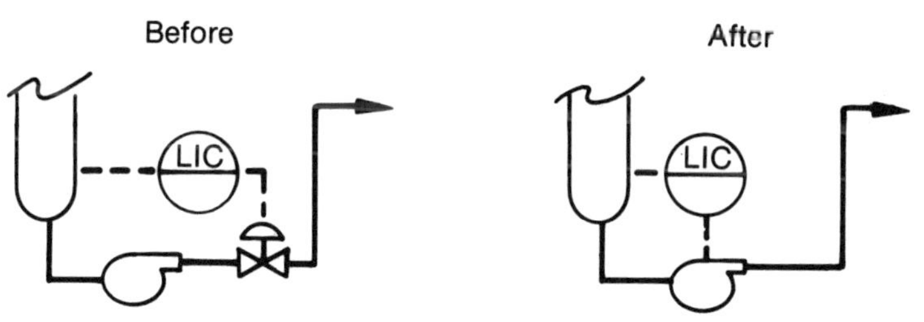

Figure 12.1a

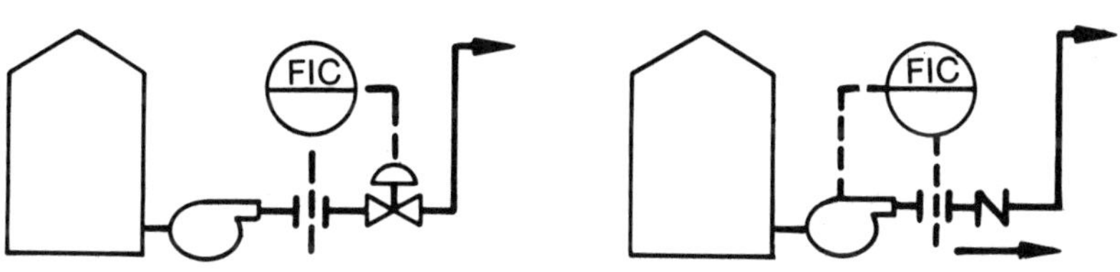

Figure 12.1b

tank. Similar to the level loop shown in Figure 12.1a, the pump generates hydraulic energy that is wastefully dissipated across the control valve. It should be noted that, in addition to process capacity requirements, the pump must be sized to overcome the additional pressure drop associated with the control valve and flowmeter at maximum flow.

The result of adding a variable speed drive to this loop is similar to that of the level loop, in that elimination of the control valve and its associated bypass piping reduce hydraulic requirements, with important special considerations. When the flowmeter can register in the reverse flow direction, it is advisable to install a check valve at the pump discharge to prevent flow reversal when the discharge pressure is insufficient to produce flow. By preventing reverse flow in the line, false signals which can result in an incorrect control action do not reach the controller.

Less hydraulic energy is required when a flowmeter exhibiting a reduced pressure drop is installed, resulting in electrical energy savings that can often justify additional expenditures to purchase the flowmeter. Savings are achieved utilizing variable speed drive technology because the equipment is throttled so as not to produce the excess energy. Note that these energy savings would not be achieved with a throttling system, as the pressure drop not taken by the low-loss flowmeter would be taken by the control valve.

Figure 12.1c illustrates a control loop in which the pressure discharged from the pumping system is reduced, thereby lowering flow (which is typical of a cooling tower pump). Similar to the level and flow loops previously presented, the pump generates hydraulic energy that is wastefully dissipated across the manually adjusted valve. The discovery of the application of this control strategy is made in the field by examination of the pump discharge (shutoff) valving under various operating conditions. It should be clearly understood that this control strategy is very attractive when one considers that an operational problem can be resolved in the field with no technical consultation or capital investment. Electrical energy costs are actually reduced somewhat by this control strategy, since, by reducing flow, motor load and hence electrical costs are reduced.

Utilization of a variable speed drive can further reduce the electrical energy requirements of the pumping system by reducing the amount of hydraulic energy actually produced. Control strategies that match pump speed to all of the loads associated with multiple users are usually complicated and capital intensive. As a trade-off in most multiple user systems, pump speed can be approximately matched to load by maintaining a constant pump discharge pressure. When a user requires more (less) liquid, the header pressure will fall (rise), thereby increasing (decreasing) pump speed to meet demand.

Figure 12.1d shows a control loop in which the pressure discharged from the pumping system is reduced by bypassing flow back to the vessel, thus increasing flow through the pump. This control strategy is similar to the manually adjusted pump discharge valve example, except that the flow through the pump is the sum of the load and bypass flows. The discovery of the application of this control strategy is typically made in the field by examination of the piping system and is often difficult to locate due to the intricate nature of the piping layout. Similar to the manually adjusted pump discharge valve example, this control strategy requires virtually no technical consultation and minimal capital investment. Electrical energy costs are usually increased by this control strategy in that, as a percentage, flow is increased more than the discharge pressure is decreased, increasing motor load and, hence, electrical costs.

Electrical energy requirements of the pumping system can be reduced by utilizing pump speed to control discharge pressure. This control strategy has

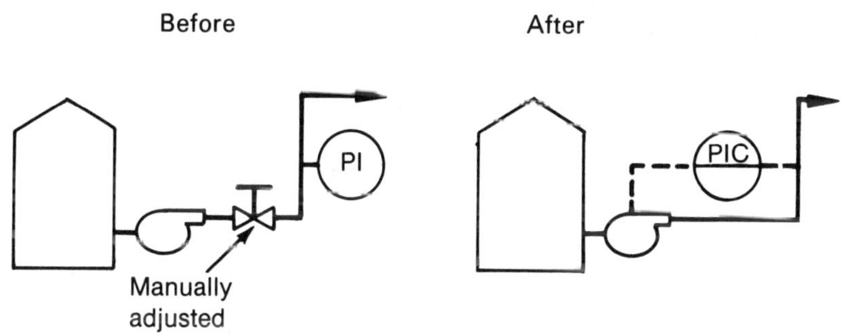

Figure 12.1c

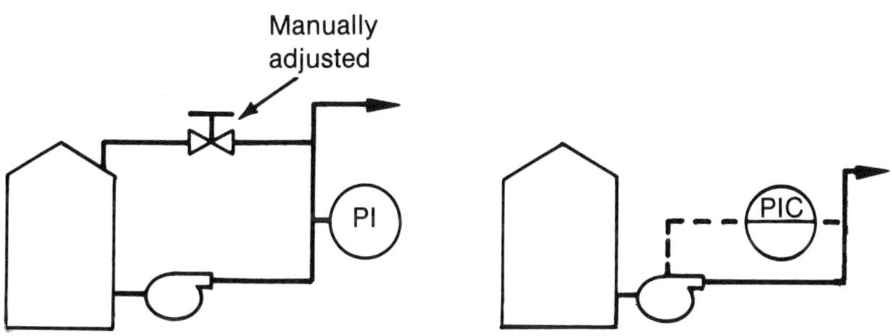

Figure 12.1d

the advantage of enabling the bypass to be completely closed, totally eliminating the hydraulic losses associated with the manual bypass valve.

Figure 12.1e illustrates a control loop in which flow of a liquid into a vessel is regulated by the temperature of the contents in the vessel, typical of an exothermic chemical reactor. The control strategy is similar to the examples above; however, one of the functions of the control valve is to provide tight shutoff. Due to operational convenience, a common practice with this control strategy is to let the motor operate continuously, whether the liquid is needed or not.

In this application, temperature is used to control the motor speed that regulates flow to the vessel. An on/off valve (which is less expensive than a control valve) must be installed due to safety considerations.

FAN/BLOWER APPLICATIONS

Speed control of a fan or blower represents a significant opportunity for the application of variable speed drive technology in new as well as retrofit applications. The economic advantages of applying variable speed drive technology to fans and blowers will be significantly greater than those attainable in a pumping application utilizing an identical motor, due to the virtual lack of static head typical of most fan and blower applications.

Figure 12.2a shows a control loop in which air flow is controlled by throttling the fan inlet damper, typical of a boiler combustion air flow control loop. Control of flow under low flow conditions is difficult due to the leakage inherent to dampers. In this application, flow is varied by varying a restriction at the fan inlet, thus generating hydraulic energy, which is wastefully dissipated across the fan damper. Note that installation of a

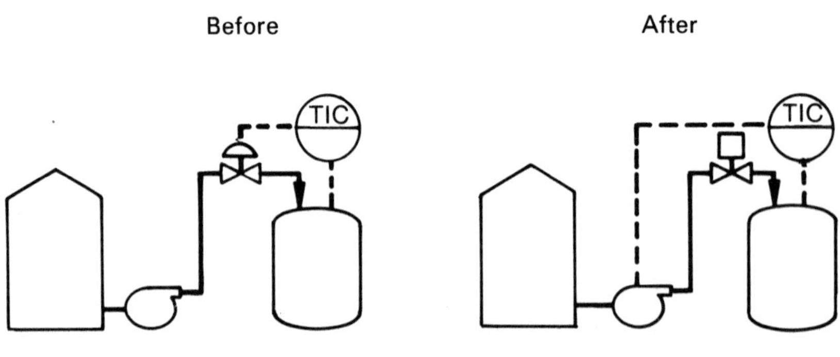

Figure 12.2e

flowmeter that operates with no significant pressure drop will *not* reduce energy consumption because the hydraulic energy not dissipated across the flowmeter is dissipated across the fan damper.

Adding a variable speed drive to this loop eliminates the fan damper (often a high maintenance item), provides superior control at low flow conditions, and reduces electrical energy costs. Installation of a flowmeter exhibiting a reduced pressure drop will result in electrical energy savings, which can often justify the additional expenditures to purchase the flowmeter. To avoid process problems, it is advisable to select a flowmeter that will produce an output only in the forward flow direction in applications where reverse flow conditions can occur.

In the control loop shown in Figure 12.2b, air flow is regulated by throttling the outlet of the fan. The ramifications of applying a variable speed drive system are similar to those of the inlet damper system.

In the control loop in Figure 12.2c, air pressure is controlled by bleeding air to the atmosphere. In this application, the blower operates at full capacity while air pressure can be varied by discharging excess air to the atmosphere, wasting hydraulic energy in the process. Electrical costs can be high since the motor remains fully loaded, incurring fully loaded motor electrical billing, independent of the actual load.

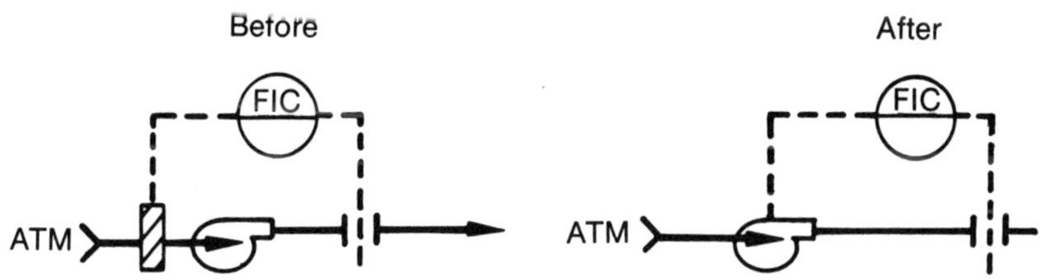

Figure 12.2a

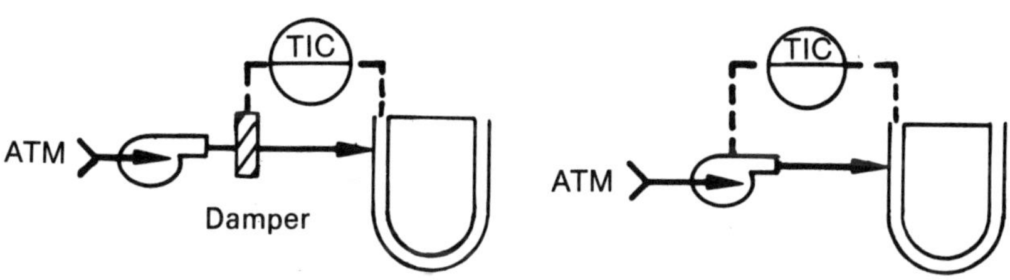

Figure 12.2b

Adding a variable speed drive to this system, allowing the pressure to regulate blower speed, eliminates the control valve, its bypass piping, and the noise problems associated with atmospheric discharge. Large electrical energy savings are usually associated with this application.

COMPRESSOR APPLICATIONS

Speed control of a compressor represents a significant economic opportunity for the application of variable speed drive technology in new as well as retrofit applications, primarily due to the size of the drive motor and the amount of electrical energy it consumes. The benefits of superior control and process improvements must be examined on an individual basis.

The ability of a compressor to function properly at reduced speed is paramount to a successful variable speed drive application. The most commonly encountered constraints at reduced speeds are inadequate lubrication and compressor inefficiency, which may cause high temperatures and excessive compressor wear. Depending on compressor design, minimum operating speeds typically vary from 0 to 50 percent of full speed.

The control loop in Figure 12.3a shows gas pressure being controlled by bypassing gas to the compressor inlet. This may be implemented through utilization of a control valve or by mechanical means within the compressor. In this application, the positive displacement compressor operates at full capacity while gas pressure is varied by bypassing excess gas (cooled) to the compressor inlet, wasting hydraulic energy. Large electrical costs are incurred because the motor remains fully loaded (incurring fully loaded motor electrical billing) independent of the actual load.

Allowing the discharge pressure to regulate compressor speed matches the load to the required gas flow, eliminating the operation of bypass devices. Bypass devices should be active only to reduce capacity when the

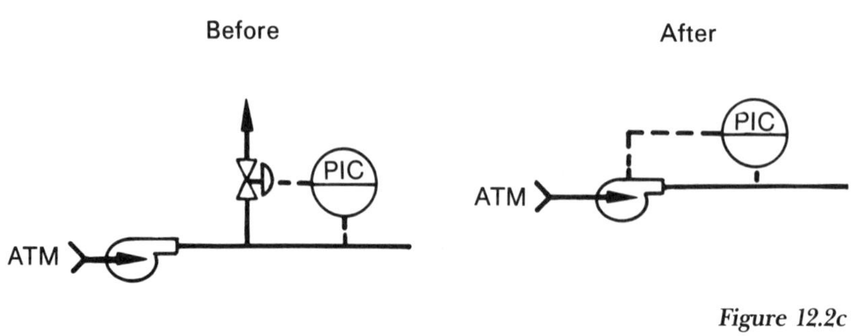

Figure 12.2c

compressor is operated at its minimum operating speed with an insufficient load. Large electrical energy savings are typically associated with this configuration due to the size of the motor.

Figure 12.3b illustrates a control loop in which gas pressure is controlled by bleeding excess gas to the atmosphere. In this application, the compressor operates at full capacity, thus wasting hydraulic energy. This can result in large electrical costs because the motor remains fully loaded (incurring fully loaded motor electrical billing) independent of the actual load, as in the previous example.

Adding a variable speed drive to this loop, allowing the pressure to regulate compressor speed, produces results similar to those of the bypass system.

A control loop in which high and low gas pressure switches control the outlet pressure of the compressor and receiver systems between minimum and maximum limits by loading and unloading the compressor is shown in Figure 12.3c and is typical of an air compressor. In this application, the compressor operates at full capacity while the compressor is loaded and at

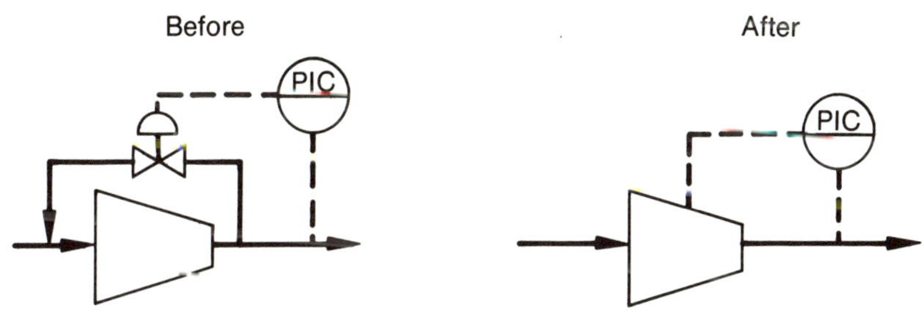

Figure 12.3a

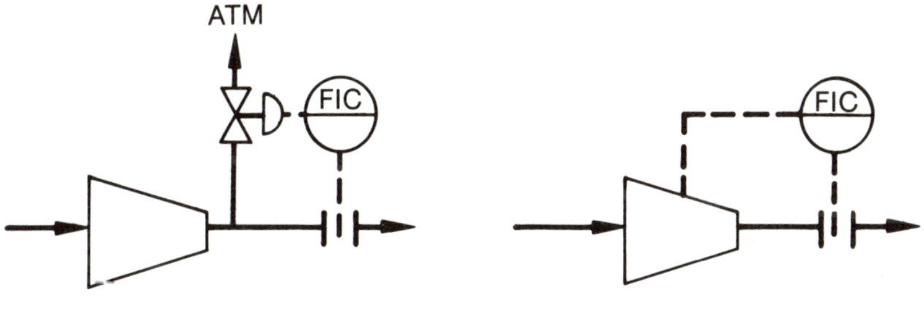

Figure 12.3b

zero capacity when the compressor is unloaded. This arrangement usually does not waste much hydraulic energy, but the motor must operate while the compressor is unloaded and performing no work, thereby wasting electrical energy. Note that some energy is wasted because the discharge pressure of the loaded compressor may be higher than system requirements.

Adding a variable speed drive to this system, allowing the discharge pressure of the system to regulate compressor speed and match the required gas flow, eliminates the wasted electrical energy associated with operating the compressor unloaded and, in some designs, the wasted energy to compress the gas to the higher pressure when loaded. Unloading devices should be active only to reduce capacity when the compressor is operated at its minimum operating speed with an insufficient load. Significant electrical energy savings are typically associated with this configuration due to the size of the motor.

Figure 12.3d shows a control loop in which the compressor suction pressure is controlled by internally bypassing gas to the compressor inlet, typical of a refrigeration compressor. This application is similar to that of the bypass valve, except that the suction pressure, rather than the discharge pressure, is controlled.

MISCELLANEOUS APPLICATIONS

Variable speed drive technology can be applied to many applications in which rotational or electrical energy input must be throttled. For example, the speed of a cooling tower fan can be varied to control cooling water

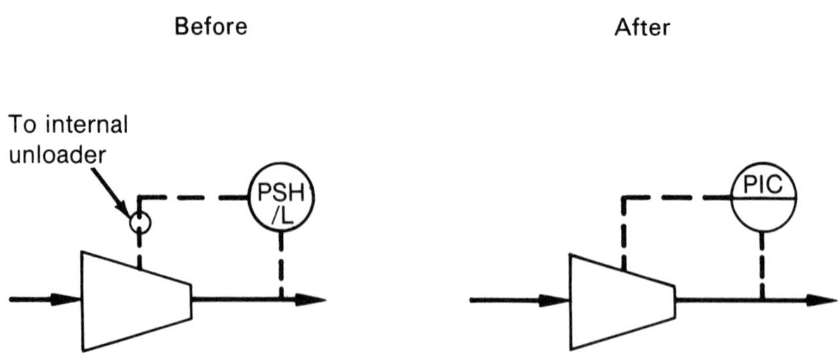

Figure 12.3c

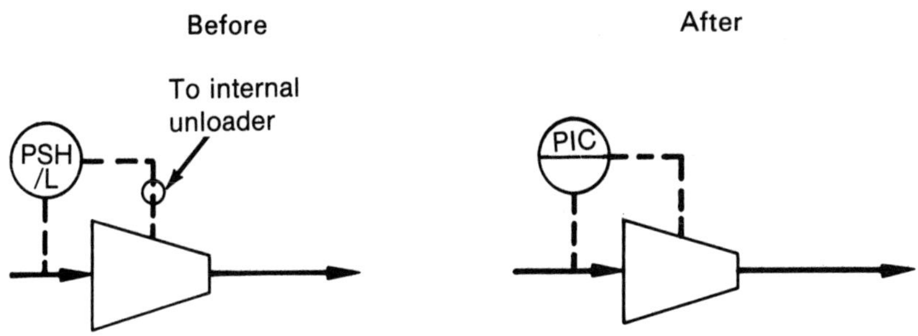

Figure 12.3d

temperature. While an electric heater can be controlled by a variable frequency drive, other technologies, such as a thermostatic control, may be more cost-effective in this application.

Each application of variable speed drive technology must be considered individually, making the nature of all applications somewhat miscellaneous.

Selling a Project: The Economic Approach

13

Given an appreciation for the advantages and disadvantages of variable speed drives, a technical person can readily identify at least one potential application that competes with a more conventional method. The temptation is to rush to present the application to one's direct supervisor, excitedly expounding on the logic and virtue of applying variable speed drive technology in the application.

While this approach may seem logical, in the overwhelming number of cases the supervisor's reaction is negative for a variety of apparent reasons, many of which are legitimate. The supervisor may have too many other more immediate problems and may not be willing to pursue such a project at the time. A lack of an adequate understanding of the concept, equipment, and/or instrumentation involved may cause the supervisor to reject the project out of hand. On a larger scale, the company may not have the financial resources or necessary manpower to implement the project.

A supervisor who is in favor of implementing a project may find it difficult, if not impossible, to sell the project to upper management on its technical merit unless variable speed drive technology is absolutely required for the application. Finally, it is unlikely that the financial persons, who often ultimately decide where resources will be allocated, will have an appreciation for the merit of the project.

The approach outlined above (that is, making emotionally logical suggestions for the application of variable speed drive technology) should be avoided to insure the best chance of implementing the project. The proper approach is to MAKE MANAGEMENT AN OFFER THAT CANNOT BE REFUSED. To enhance the chances of convincing management that the project is worthy of support, the offer should be made in terms that management understands — MONEY.

Simply stated, making management an offer that cannot be refused can be done by calculating the costs associated with the present design and comparing them to those associated with alternate designs, which may incorporate variable speed drives. The net result of these calculations is to quantify the costs (installation, operation, and maintenance) associated with various alternatives, allowing technical personnel to pursue the most viable projects. Presentation of the results of the above calculations to management greatly enhances the chances of obtaining the resources necessary to implement a project.

The economic measure and the calculations with which the viability of an expenditure is analyzed can and do vary considerably from company to company. Appropriate persons within each company are capable of performing these calculations, given the expenditure required and the savings that result. For simplicity, the simple payback, defined as

Simple Payback (years) = Expenditures/Annual Savings

is used here, although more complex measures of evaluating a project may take into account considerations such as taxes, interest rates, inflation, and the like.

ORGANIZATION OF ANALYSIS

Development of a straightforward method to analyze and present data regarding an application is difficult due to the varied nature of the applications of variable speed drives. However, some guidelines can be applied to most applications, where various design alternatives are to be compared on a common basis.

(1) The various designs should be itemized as different scenarios, the first scenario typically being the present design, either proposed (for new construction) or of an existing installation.

(2) The installation, operation, and maintenance costs should be calculated for each scenario independently, duplicating calculations when necessary. This is important due to confusion that can occur, especially when management requests an evaluation of four or more alternative scenarios.

(3) The savings, expenditures, and simple payback for each scenario, as compared to one scenario (typically the present design), should be calculated.

(4) The results should be economically evaluated.

SCENARIO DEVELOPMENT

As a minimum, the present design and the proposed design must be defined for comparison purposes. The originator of the alternate concept will usually develop one or two alternate scenarios to the proposed design, but a few others will probably be suggested by management.

Each scenario should be defined in sufficient detail, either in writing or utilizing sketches, to perform the calculations outlined below. All assumptions should be documented in writing to enhance clarity and allow verification of the design basis, while all supporting documentation, such as pump curves and the like, should be accumulated. It is not uncommon to develop four or more scenarios for a typical application.

It is usually convenient to select one scenario, typically the present design, to be the *base* scenario, that is, the scenario to which all other scenarios are compared.

INSTALLATION COSTS

The installation costs associated with a scenario include the costs of *all items necessary to effect a complete operating system*, some of which are difficult to quantify. The installation costs for each scenario should be calculated either by summing all of the costs associated with the application or by variance from the base scenario. Calculations for retrofit applications involve the calculation of all costs necessary to convert the existing design to the proposed designs.

The installation costs associated with the application of a control valve final control element (which is often the base scenario) include the valve, positioner, regulator, current-to-pneumatic converter, bypass valves, piping and fittings, instrument air supply piping and valves, conduit and fittings, field cables, engineering, and labor to install all of the above. In some applications, other installation costs may be applicable and should not be neglected, such as additional tracing and insulation costs due to the intricate nature of the valve manifold, costs for additional walkways and gratings to provide access to the valve manifold, the cost of purchasing and installing a combination starter, and the like. Because each application is installed using different design criteria with varying degrees of proximity to utilities, each should be considered on an individual basis.

The installation costs associated with a variable speed drive as the final control element include the drive and the costs associated with mounting it. The cost of providing a suitable location to mount the drive may be significant, such as when there is insufficient space in an existing motor control center or when it must be installed between the existing motor and the equipment. The costs associated with instrumentation not included in the base scenario, typically a transmitter, controller, conduit and fittings, engineering and labor to install the above, should not be neglected.

Note that the common costs, such as the cost of running the conduit and cables to the motor and push button, are not considered as they are required in both the control valve and variable speed drive scenarios. It is often easier to use the approach of calculating the cost *differences* between scenarios, but care must be taken to avoid confusion by calculating the cost *differential* between each scenario and the base scenario.

OPERATING COSTS

Operating costs should be calculated for each scenario; although operating cost differences may be more conveniently calculated in some instances. In typical variable speed drive applications, operating cost savings can be achieved, but it should be noted that savings due to modification of the process can be considerably larger than electrical cost savings. Therefore, variable speed drive projects that yield process as well as electrical savings are usually very attractive.

Regardless of the type of equipment, operating cost calculations should weigh the amount of time equipment is operated at each load. For example, equipment operated at full load 24 hours per day, 365 days per year will have a significantly higher operating cost than the same equipment operated for 40 hours per week. Data concerning the amount of time equipment is operated at each load is usually difficult to obtain, but reasonable estimates can be made by those familiar with the operation.

Cost calculations should utilize measurements or equipment performance curve data; however, an approximate model of equipment operation may be necessary, when specific equipment data is not available, in order to produce an adequate approximation of operating costs.

The following sections provide some guidelines for the calculation of the operating costs associated with equipment operating at different load conditions. The operating costs associated with each scenario should be calculated for each load condition and weighed accordingly to calculate the annual operating cost associated with the equipment.

The overall strategy is to calculate the required discharge pressure and flow from which the brake horsepower, equipment efficiency, and operating costs can then be calculated. The operating costs associated with process considerations should be evaluated on an individual basis.

Discharge Pressure of Constant Torque Equipment

Constant torque equipment is typified by positive displacement pumps and compressors, but other types of equipment may be included as well. At full speed, the flow and discharge pressure are virtually constant due to the fixed compression ratio and displacement of the equipment. The net result is a brake horsepower requirement that is equal to the full load brake horsepower for virtually all system operating conditions.

Neglecting slippage, the attainable discharge pressure of positive displacement equipment remains constant at all speeds, while its flow is directly proportional to the speed at which it is operated. As a result, the brake horsepower can be estimated as

BHP = BHP at Full Speed × Percent Speed

It should be noted that some positive displacement equipment may not be legitimately modeled as constant torque devices due to internal mechanical adjustments or other anomalies that may alter the constant torque behavior of the equipment. These devices should be considered on an individual basis. However, when the control strategy is such that motor speed is varied to maintain fully loaded equipment operation, the constant torque model can often be applied.

Discharge Pressure of Centrifugal Equipment

Centrifugal equipment is typified by centrifugal fans, blowers, and pumps but may include other types of equipment as well. While the brake horsepower requirements may be read directly from operating curves with reasonable accuracy, equations that estimate brake horsepower may be more useful for calculation purposes.

Figure 13.1 shows the operation of a typical piece of centrifugal equipment, where the discharge pressure at no flow conditions and the flow,

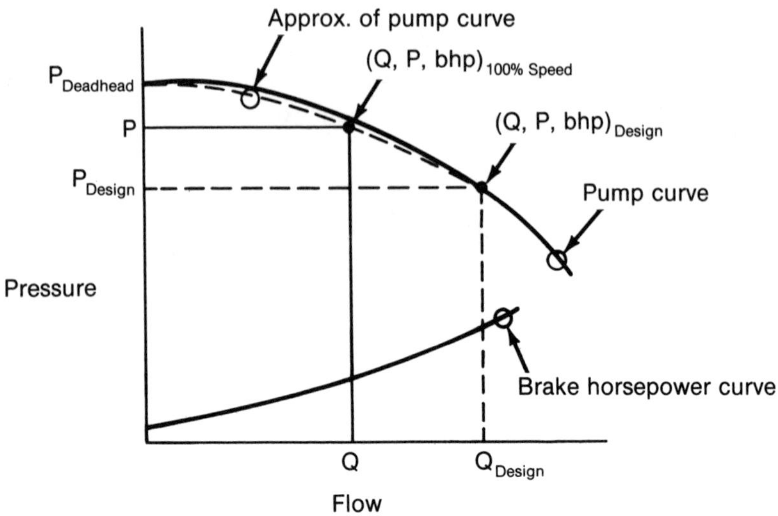

Figure 13.1
Full Speed Operation of Centrifugal Equipment

discharge pressure, and brake horsepower at design conditions can be readily determined. By using a second-order (quadratic) equation to estimate equipment curve, the discharge pressure at a given flow can be estimated by

$$P_{100\%\text{ speed}} = P_{\text{Design}} + (P_{\text{Deadhead}} - P_{\text{Design}}) \times \sqrt{1 - \frac{Q}{Q_{\text{Design}}}}$$

It should be noted that while decreasing flow causes an increase in discharge pressure, some applications, notably blowers, can exhibit the opposite response at lower flows. However, the equation should provide an adequate estimate of discharge pressure for most constant speed applications.

Variations in equipment speed cause the flow and discharge pressures of centrifugal equipment to be governed by the Affinity Laws, where a third-order (cubic) relationship exists between speed and energy consumption.

$$\text{BHP} = \text{Full Load BHP} \times (\text{Percent Speed}/100)^3$$

Therefore, speed reduction of centrifugal equipment shows significant energy-saving potential and can be likened to varying impellor size.

The operating curves of a typical piece of centrifugal equipment, where the flow, discharge pressure, and brake horsepower at full speed design conditions can be readily determined, are shown in Figure 13.2. When sufficient information is available, the system curve can be plotted, and the required system pressure at a given flow can be determined graphically. As this information is typically not available, the required system pressure can be estimated from the full speed design conditions and piping considerations by making a few approximations.

As the operating curve illustrates, equipment designed to operate at full speed is sized to overcome the static head of the piping system, piping friction losses, and the pressure drop associated with the control valve before the discharge pressure requirements are increased (by design) to provide a margin of excess pumping capacity. The required discharge pressure at design flow is given by

$$P_{\text{Required}} = P_{\text{Design}}\left[1 - \left(\frac{P_{\text{Design}} - P_{\text{Cv}}}{P_{\text{Design}}}\right) - \left(\frac{P_{\text{Cv}} - P_{\text{Required}}}{P_{\text{Cv}} - P_{\text{Static}}}\right)\left(\frac{P_{\text{Cv}} - P_{\text{Static}}}{\text{Design}}\right)\right]$$

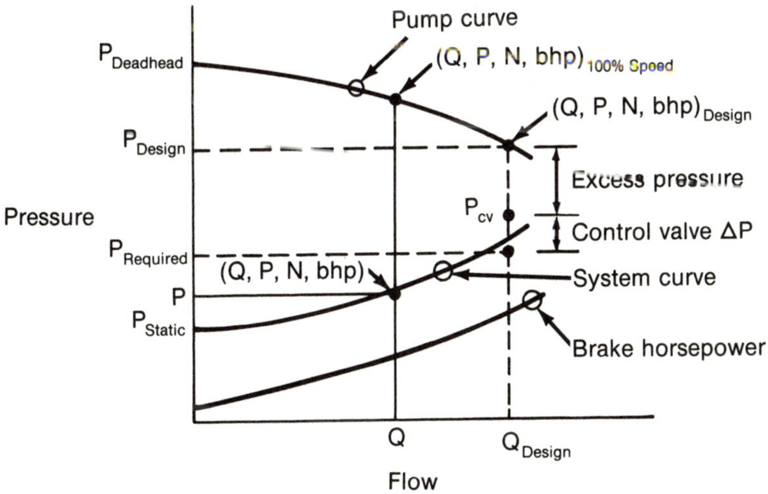

Figure 13.2
Variable Speed Operation of Centrifugal Equipment

where the excess pumping capacity is given by

$$F_{\text{XS cap}} = \left(\frac{P_{\text{Design}} - P_{\text{Cv}}}{P_{\text{Design}}}\right) = \left(1 - \frac{P_{\text{Cv}}}{P_{\text{Design}}}\right)$$

The ratio of the differential pressure across the control valve to the piping friction losses is

$$F_{\text{Cv}} = \left(\frac{P_{\text{Cv}} - P_{\text{Required}}}{P_{\text{Cv}} - P_{\text{Static}}}\right)$$

while the ratio of static pressure requirements to design pressure is given by

$$F_{\text{Static}} = \frac{P_{\text{Static}}}{P_{\text{Design}}}$$

The expression for the required discharge pressure can be rewritten as

$$P_{\text{Required}} = P_{\text{Design}}\,[1 - F_{\text{XS cap}} - F_{\text{Cv}}\,(1 - F_{\text{XS cap}} - F_{\text{Static}})]$$

where the required discharge pressure at design flow is a function of the full speed design discharge pressure, static head requirements, excess pumping capacity, and the percentage control valve differential pressure.

In a typical application, the pump is oversized by approximately 15 percent, and the pressure drop across the control valve is approximately 30 percent of the piping friction losses to provide adequate control. Using these assumptions,

$$P_{\text{Required}} = P_{\text{Design}}\left[1 - (0.15) - (0.30)\left(1 - 0.30 - \frac{P_{\text{Static}}}{P_{\text{Design}}}\right)\right]$$

$$= 0.64\ P_{\text{Design}} + 0.30\ P_{\text{Static}}$$

Note that these parameters may differ in actual applications.

A second-order (quadratic) equation can be used to estimate the pressure at flow Q as

$$P = P_{\text{Static}} + \left(\frac{Q}{Q_{\text{Design}}}\right)^2 (P_{\text{Required}} - P_{\text{Static}})$$

Using the Affinity Laws, the equipment speed is

$$\left(\frac{N}{N_{\text{Design}}}\right) = \left(\frac{P}{P_{100\%\ \text{Speed}}}\right)^{1/2}$$

assuming that the equipment efficiency remains constant as speed is varied. This assumption is reasonable in most applications.

Brake Horsepower

Utilizing operating data, equipment performance curves, or the above calculations, the brake horsepower at reduced speed can be estimated using

$$bhp = \left(\frac{N}{N_{Design}}\right)^3 bhp_{Design}$$

assuming that equipment efficiency remains relatively constant as speed is decreased.

Efficiency

Equipment efficiency is the ratio of the hydraulic energy produced to the brake horsepower required where

Equipment Efficiency = Hydraulic Energy/BHP

A perhaps more meaningful measure of operating efficiency is the ratio of the billed costs to the useful hydraulic energy output of the system.

Billed Electrical Costs

For full speed and reduced speed applications, the billed electrical energy consumption can be estimated as

$$\left[\frac{(0.15 + \Delta\eta) + (0.85 - \Delta\eta)\left(\frac{\text{Actual Load}}{\text{Motor Capacity}}\right)}{(0.93 - \Delta\eta)}\right](\text{Motor Capacity})$$

Table 2.1

Motor Capacity		$\Delta\eta$
0–3 hp	0–2.2 kW	0.18
5–10	3.7–7.5	0.15
15–25	11.2–18.7	0.12
30–40	22.4–44.8	0.09
50–60	37.3–44.8	0.06
75–100	56.0–74.6	0.03
over 125	over 93.3	0.00

where

Motor Capacity = Full Load Capacity × Percent Speed/100

kVA demand can be estimated using

$$\text{Current} = \left[0.35 + 0.65\left(\frac{\text{Actual Load}}{\text{Motor Capacity}}\right)\right] \text{(FLA)}$$

Electronic variable speed drive losses are typically less than 4 percent of motor capacity, while kVA demand is lower and distribution is slightly more efficient compared to constant speed operation due to the higher power factor. As a result of the small and somewhat offsetting effects of drive losses and increased distribution efficiency, the net effect of these phenomena are small and are usually neglected.

From the above discussion, billed electrical consumption can be calculated as the product of the energy requirement and the incremental energy cost. In some applications, calculation of the kVA demand charges and distribution losses may be necessary to justify the purchase of power factor correction equipment.

By estimating the billed electrical consumption for various load conditions and the amount of time during which the equipment is operated at each load condition, electrical operating costs can be adequately estimated.

Process Considerations

In some applications, better control or process modification made possible by variable speed drive technology may result in handsome process savings. Due to the customized nature of each process, guidelines for identifying possible process cost savings are intentionally vague.

Primary to investigation of potential process considerations is an understanding of the process to be evaluated and access to a person who is intimately involved with the operation. It should be recognized that the sum of raw materials, labor and energy usually represents the great majority of the operating costs associated with the process. Therefore, where utilization of variable speed drive technology affects any of these parameters, potential process savings may exist.

For example, a variable speed drive that operates a cooling fan continuously (as opposed to cycling on/off) may stabilize the process, resulting in cost savings. A variable speed drive that provides better control of one raw material may result in the usage of less of another raw material, while still maintaining product quality.

MAINTENANCE COSTS

The annual maintenance cost associated with a variable speed drive application is usually difficult to quantify unless the drive represents the solution to a known maintenance problem. For example, the amount of maintenance required to repair a control valve with continual leakage problems can be quantified, but the reduction in the frequency with which a piece of equipment is overhauled as a result of slower operation is usually difficult to quantify when years pass between consecutive overhauls.

While difficult to quantify, maintenance associated with a electronic variable speed drive installation is probably less than that of a control valve installation. However, variable speed drives that utilize other technologies can require substantially more maintenance than its equivalent control valve system.

NET OPERATING COSTS

The operating cost associated with each scenario can be calculated by summing the effects of electrical operating cost calculations, process considerations, and maintenance cost calculations.

ECONOMIC EVALUATION

Having calculated the net operating cost associated with each scenario, the simple payback may be calculated as the ratio of total expenditures to the total savings obtained when compared to the base scenario. From the simple payback calculation, the potential of each scenario can be quickly evaluated to determine if the project should be pursued and a more detailed technical analysis performed.

It should be noted that the economic viability of a project is only one indication of the importance. In some applications, variable speed drive technology may provide the means to accomplish an important end that is not quantifiable, such as controlling a difficult loop or meeting environmental standards. In these applications, electrical energy savings are incidental to the project and only add to its attractiveness.

Applications

14

Both new and retrofit installations provide a substantial number of viable variable speed drive applications. To achieve maximum benefit from the technology, each application should be viewed as a modification of the total system rather than as an alteration to a piece of process equipment.

The examples presented below are intended to illustrate the concepts used to evaluate some potential applications of variable speed drive technology. Of importance are the techniques with which the evaluation is performed and the significant percentage of electrical energy savings that can be achieved through the use of variable speed drives.

In the process of applying technology one must avoid the temptation to get involved in an unnecessary or economically unsound project. For example, varying the speed of undersized equipment serves no purpose, since the equipment will virtually always operate at 100 percent of speed. While variable speed drive technology should be applied where they make economic sense, they should not be applied in a wholesale manner without proper evaluation.

Certain details are intentionally deleted from the ensuing analysis to allow the user to focus on major improvements. For example, in the great majority of viable applications, the power factor penalty, drive efficiency, and electrical distribution costs can be neglected because the savings associated with the process and equipment operation are usually an order of magnitude larger than their combined effects.

For clarity, the examples below illustrate applications where the motor operates continuously (8760 hours/year). Times during which the equipment actually operates (both the amount of time and the time of day) can have a profound effect on the viability of an application.

Equipment and labor costs presented here are primarily intended to illustrate concepts; they can and will vary significantly with manufacturer, location, and facility. The applications described below are typical and any resemblance to actual applications is incidental.

SINGLE-USER CENTRIFUGAL PUMPING SYSTEM

Pumping systems are used extensively throughout industry to transport liquids to a single user, such as a reactor, or to some other piece of process equipment.

Figure 14.1 illustrates a single-user centrifugal pumping system in which a control valve is used to dissipate hydraulic energy in order to control flow to the user.

In many applications, the operating flow can be approximately 60 percent or less of the design capacity of the pumping system. This is due to a tendency to set the range of the flowmeter 25 percent above the maximum flow requirements (to ensure that flow rates do not exceed the flowmeter range) before an additional 15 percent excess pumping capacity is added to the system.

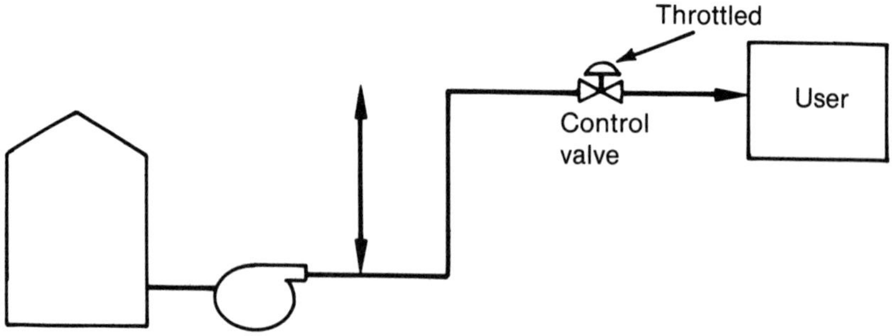

Figure 14.1
Single-User System

In applications where the actual flow is less than design flow, the control valve must dissipate significant amounts of hydraulic energy to maintain the desired flow. Utilizing a variable speed drive to generate only the required hydraulic energy for the load will reduce energy consumption and eliminate the control valve and bypass piping.

Scenario I *Water Flow Control Loop (Existing)*

A typical flow control loop in which water flow to a user is controlled by throttling a control valve is shown in Figure 14.2a. The pump model number was taken from the pump tag; impeller information was verified in the plant maintenance files; manufacturer literature was used to locate the pump curve shown in Figure 14.2b.

The pump has a 6½ inch (165 mm) impeller, is designed to produce 200 gpm (45.5 m^3/hr) of flow at a discharge pressure of 165 feet (50.3 meters), and requires approximately 14.0 brake horsepower (10.4 kW). The pump is driven continuously by a 15 hp (11.2 kW) motor. The actual flow to the user is controlled at 125 gpm. Note that the pressure inside the vessel is 10 psig (23.1 feet WC) and that the elevation of the user is 50 feet above the pump.

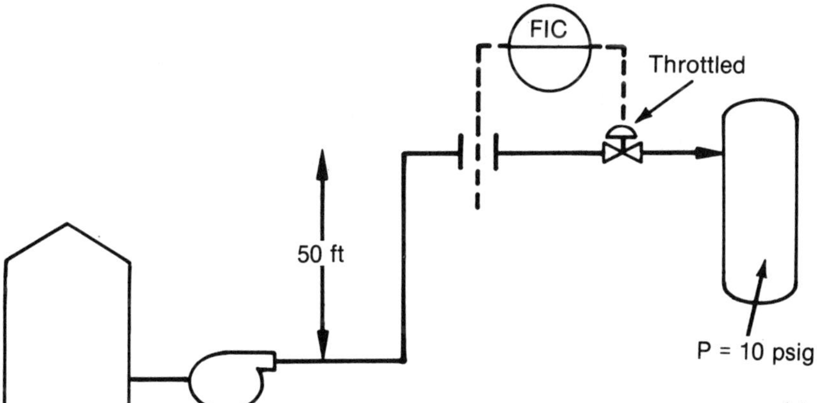

(a)

Figure 14.2
Scenario I — Water Flow Control Loop

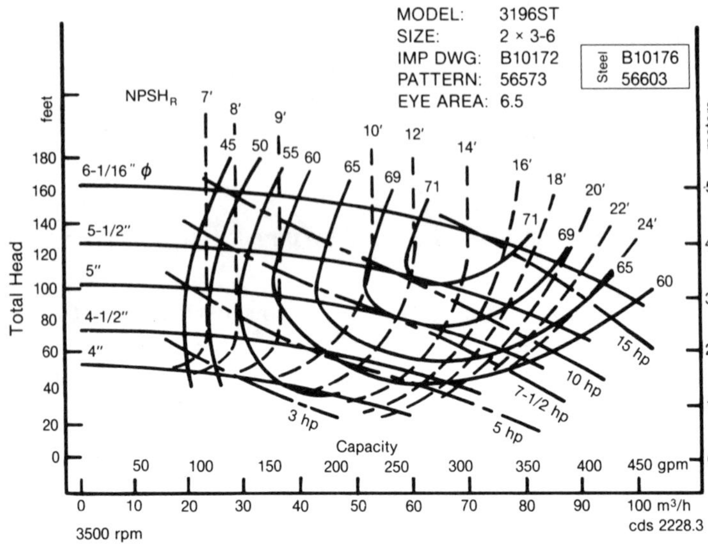

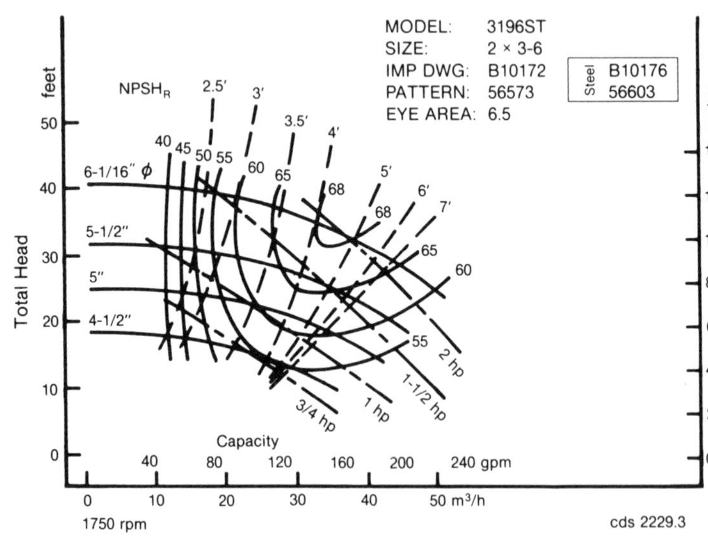

Figure 14.2 (continued)

The pump discharge pressure at actual flowing conditions can be determined from the pump curves as approximately 190 feet, or estimated as

$$P_{\text{Design}} = 165 \text{ feet}$$

$$P_{\text{Deadhead}} = 200 \text{ feet}$$

$$Q = 125 \text{ gpm}$$

$$Q_{\text{Design}} = 200 \text{ gpm}$$

$$P_{100\%\text{ Speed}} = P_{\text{Design}} + (P_{\text{Deadhead}} - P_{\text{Design}})\sqrt{\left(1 - \frac{Q}{Q_{\text{Design}}}\right)}$$

$$= 165 \times (200 - 165)\sqrt{\left(1 - \frac{125}{200}\right)}$$

$$= 186 \text{ feet}$$

Therefore, the control valve and piping must dissipate (186 – 50 – 23.1), or 112.9 feet WC (48.9 psi). As piping losses are typically small at this reduced flow, the control valve must dissipate the great majority of the energy. The pump curve can be used to estimate the brake horsepower requirement at 125 gpm, which is approximately 11.0 hp, from which energy consumption can be estimated as

$$\Delta\eta = 0.12$$

$$\text{Actual load} = 11.0 \text{ hp}$$

$$\text{Motor capacity} = 15.0 \text{ hp}$$

$$\text{Power consumption} = \left[\frac{(0.15 + \Delta\eta) + (0.85 - \Delta\eta)\left(\dfrac{\text{Actual load}}{\text{Motor capacity}}\right)}{(0.93 - \Delta\eta)}\right] \times \text{Motor capacity}$$

$$= \left[\frac{(0.15 + 0.12) + (0.85 - 0.12)\left(\dfrac{11.0 \text{ hp}}{15.0 \text{ hp}}\right)}{(0.93 - 0.12)}\right] \times (15.0 \text{ hp})\left(\frac{0.746 \text{ kW}}{\text{hp}}\right)$$

$$= 11.1 \text{ kW}$$

The annual operating cost of this system is approximately 11.1 kW × US$539.36/kW·yr, or US$5987.

Scenario II
Water Flow Loop with Variable Speed Drive

Figure 14.3 illustrates the proposed system in which the flow control loop will vary pump speed, eliminating the control valve. As the hydraulic energy generated will follow the load instead of exceeding the load (as in Scenario I), energy requirements will be less than in the existing system.

The pump discharge pressure at design flow can be determined by a detailed calculation of piping system losses or estimated using

$$P_{Required} = P_{Design}\,[1 - F_{XS\ cap} - F_{Cv}\,(1 - F_{XS\ cap} - F_{Static})]$$

Assuming that the pumping system is designed for 15 percent excess pressure at design flow and the control valve in Scenario I is designed to dissipate 30 percent of the piping losses at design flow,

$$P_{Design} = 165 \text{ feet}$$

$$P_{Static} = 50 + 23.1 = 73.1 \text{ feet}$$

$$\begin{aligned} P_{Required} &= 0.64\ P_{Design} - 0.30\ P_{Static} \\ &= (0.64)\ (165 \text{ feet}) - (0.30)\ (73.1 \text{ feet}) \\ &= 83.7 \text{ feet} \end{aligned}$$

The pump discharge pressure at 125 gpm can be estimated as

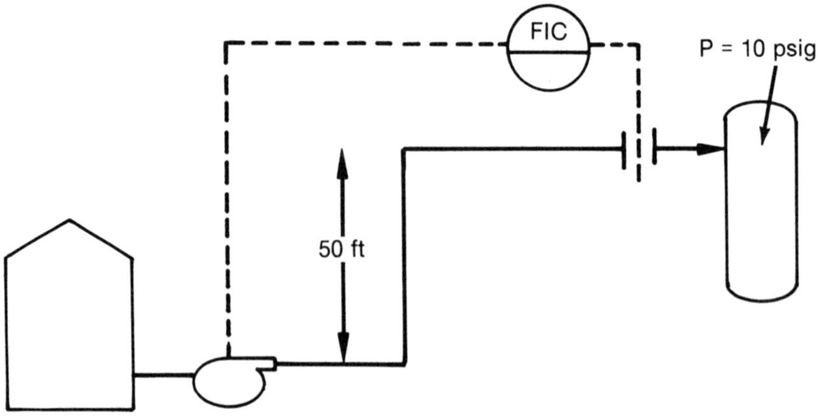

Figure 14.3
Water Flow Loop (Proposed)

$$P = P_{\text{Static}} + \left(\frac{Q}{Q_{\text{Design}}}\right)^2 (P_{\text{Required}} - P_{\text{Static}})$$

$$= 73.1 + \left(\frac{125}{200}\right)^2 (83.7 - 73.1)$$

$$= 77.2 \text{ feet}$$

The operating speed of the pump is

$$\frac{N}{N_{\text{Design}}} = \left(\frac{P}{P_{100\%\ \text{Speed}}}\right)^{1/2}$$

$$= \left(\frac{77.2 \text{ feet}}{186 \text{ feet}}\right)^{1/2}$$

$$= 64.4 \text{ percent}$$

The brake horsepower requirements are

$$\text{Brake horsepower} = \left(\frac{N}{N_{\text{Design}}}\right)^3 \text{bhp}_{\text{Design}}$$

$$= (0.644)^3 (14.0 \text{ bhp})$$

$$= 3.7 \text{ hp}$$

Power consumption can be estimated as

$$\Delta\eta = 0.12$$

$$\text{Actual load} = 3.7 \text{ hp}$$

$$\text{Motor capacity} = (0.644)(15.0 \text{ hp}) = 9.7 \text{ hp}$$

$$\text{Power consumption} = \left[\frac{(0.15 + \Delta\eta) + (0.85 - \Delta\eta)\left(\dfrac{\text{Actual load}}{\text{Motor capacity}}\right)}{(0.93 - \Delta\eta)}\right] \times \text{Motor capacity}$$

$$= \left[\frac{(0.15 + 0.12) + (0.85 - 0.12)\left(\dfrac{3.7 \text{ hp}}{9.7 \text{ hp}}\right)}{(0.93 - 0.12)}\right] \times (9.7 \text{ hp}) \left(\frac{0.746 \text{ kW}}{\text{hp}}\right)$$

$$= 4.9 \text{ kW}$$

The annual operating cost of this system is approximately 4.9 kW × US$539.36/kW·yr, or US$2643.

Alteration of the existing system can be achieved by installing an electronic variable frequency drive, the installed cost of which is estimated to be US$4000. It is assumed that the salvage value of the control valve and bypass will cover the cost of removal. It should be noted that addition of the drive will increase maintenance costs, while maintenance for the motor and pump typically decrease; however, the amounts are difficult to quantify. Maintenance associated with the control valve will also be eliminated.

Water Flow Loop Analysis

The scenarios presented above can be summarized on the following table for analysis.

Scenario	Installed Cost	Operating Cost	Savings
I	US$ 0	US$5987	US$ 0
II	US$4000	US$2643	US$3344

From the above, it can be seen that Scenario II offers a simple payback of 4000/(5987 – 2643), or 1.2 years. The percentage of reduction of electrical load is (11.1 – 4.9)/11.1, or 55.9 percent, illustrating the potential of this technology to reduce electrical energy consumption.

Had variable speed drive technology been applied in the original design, the cost of installing the control valve and bypass (approximately US$4500) would have been avoided. Also note that the installed cost of the electronic variable frequency drive typically is approximately the same or less than that of the control valve system in smaller motor sizes, often adding to the attractiveness of the drive. Typically, new installations are more attractive than retrofit applications due to lower installed costs in these sizes.

MULTIPLE-USER CENTRIFUGAL PUMPING SYSTEM

Multiple-user systems can be configured as a distribution system where the liquid is consumed by the users, such as a make-up water system, or as a circulating system where the liquid is returned to the pump and again distributed to the users, such as a cooling tower system (See Figure 14.4).

Typically, the pump and motor are sized to provide sufficient flow and pressure for the system to supply the maximum required flow simultaneously to all users. Under normal operating conditions (that is, the great majority of time during which some users are off line and others are

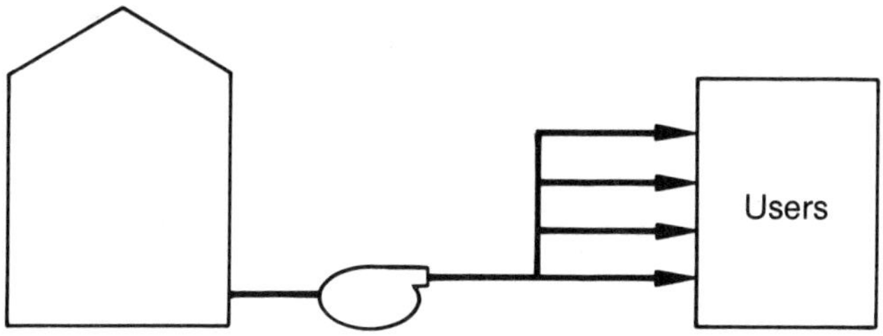

(a) Distribution System

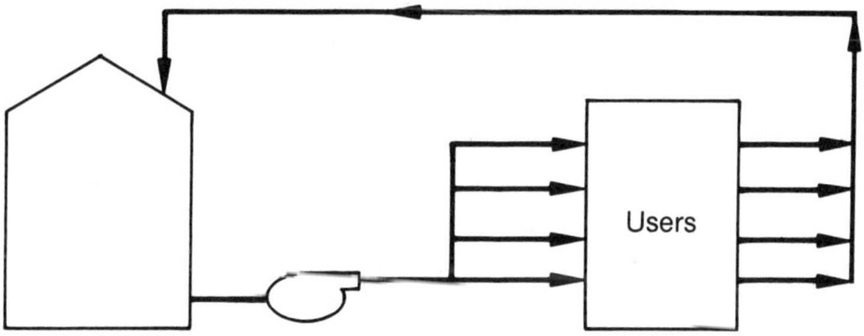

(b) Circulating System

Figure 14.4
Multiple-User Systems

operating at low flow rates), system flow and pressure requirements are typically a fraction of peak requirements. Due to the nature of the centrifugal pump curve, a centrifugal pump operating at full speed and reduced flow produces a higher discharge pressure than is required during peak operation.

Therefore, using variable speed drive technology to reduce pump speed to better match the hydraulic energy produced to load can result in significant electrical energy savings. This can be implemented by installing a control loop that controls the pump discharge pressure by varying pump speed. While this control strategy will not exactly match the hydraulic energy produced and the load, the strategy is straightforward and can be utilized to achieve sizable reductions in energy consumption without the necessity of complex control strategies and multiple control loops.

While the flow to each user in a distribution system is a function of user demand, the flow in a circulating system can often be reduced since header pressure control allows improved balancing of flows throughout the system under varying operating conditions. Thus, better system adjustment allows flow to more nearly match load and operating constraints. A reduction in flow allows a further reduction in energy consumption.

Scenario I
Circulating Water System (Existing)

Figure 14.5 is a field sketch of the existing cooling tower circulating water system. The pump curve shown in Figure 14.6 was found in the maintenance records.

The pump has a 13 inch (330 mm) impeller and is designed to produce 1800 gpm (410 m^3/hr) of flow at a discharge pressure of 150 feet (45.7 meters). The pump is driven by a 100 hp (74.6 kW) motor.

Plant operators adjust the pump discharge valve as necessary to adjust overall flow to the process. A rough determination of system flow can be obtained in the field using a portable flowmeter or by relating horsepower and/or discharge pressure to the pump curves. Discussions with operators concerning flow and the prevailing operating conditions in conjunction with field data can yield a reasonable estimate of system requirements.

In this application, flow and pressure measurements can be used to calculate the energy dissipated across the manually throttled discharge valve. For example, assuming that field measurements show that the flow, pump dis-

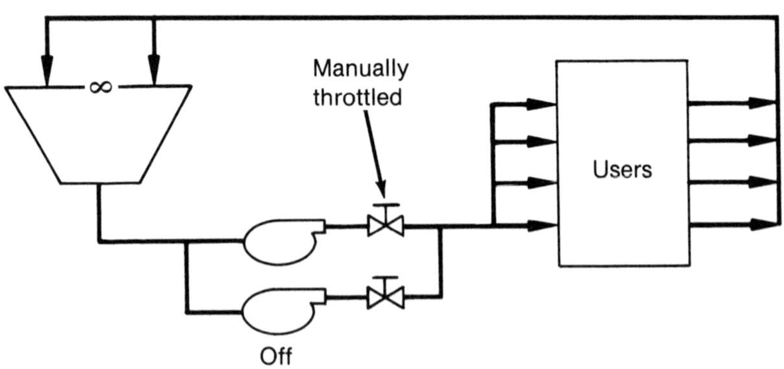

Figure 14.5
Scenario I — Circulating Water System (Existing)

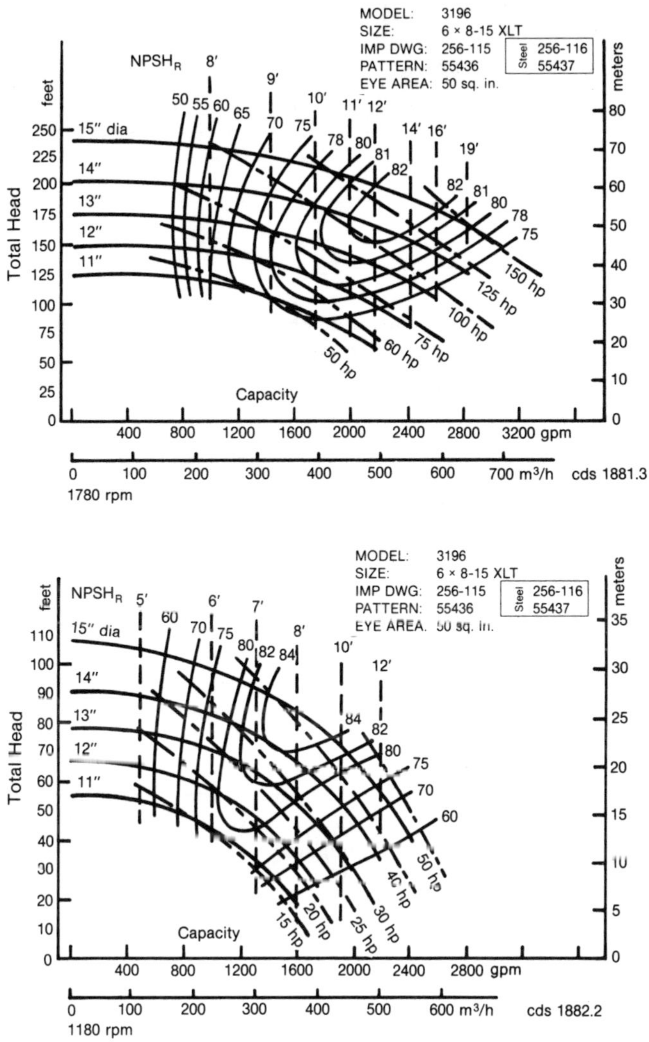

Figure 14.6
Circulating Water Pump Curve

(courtesy Goulds Pump)

charge pressure and system pressure are 1400 gpm, 70 psig, and 45 psig, respectively, the hydraulic energy dissipated across the throttled valve can be calculated as

$$\text{Wasted energy} = \frac{Q_{\text{gpm}}\,\Delta P_{\text{ft}}\,\text{SG}}{3960}\ \text{Horsepower}$$

$$= \frac{(1500\ \text{gal/min})\,(70-45\ \text{psi})\,(2.31\ \text{ft/psi})\,(1.0)\,(0.746\ \text{kW/hp})}{3960}$$

$$= 16.3\ \text{kW}$$

Due to pump and motor inefficiencies, the cost of electrical energy required to generate this amount of hydraulic energy is approximately 16.3 kW × 539.36/kW·yr/0.50, or US$17,583 per year. In larger systems, hundreds of thousands of dollars of energy can be wasted annually as a result of such throttling.

Electrical energy costs associated with operating the existing system can be calculated and summarized as shown in Table 14.1. It can be seen that the electrical energy cost associated with the operation of this motor is significant; a substantial reduction in hydraulic requirements can yield significant electrical energy savings; a 22% reduction of flow from 1800 to 1400 gpm results in only an 8% reduction in power consumption.

Table 14.1
Scenario I
Circulating Water System

Flow[1] (gpm)	Actual Motor Load[2] (bhp)	Power Consumption[3] (kW)	Operating Time[1] (%)	Weighted Power Consumption[4] (kW)
0	0.0	0.0	5	0.0
1400	75.0	65.9	15	9.9
1500	77.0	67.3	50	33.6
1600	79.0	68.6	15	10.3
1700	81.0	70.0	10	7.0
1800	83.0	71.3	5	3.6
		Total	100	64.4
		Annual Cost/kW		539.36
		Annual Operating Cost		34718

1. Estimated.
2. From pump curve.
3. $V = I \times R$, Motor Capacity = 100 hp, $\Delta\eta = 0.03$.
4. Power Consumption × Operating Time / 100.

Scenario II Circulating Water System with Variable Speed Drive

A pump discharge pressure control loop is proposed, illustrated in Figure 14.7, that will vary pump speed utilizing an electronic variable frequency drive to maintain the system pressure.

It is estimated that, due to better header pressure control, the total circulating water flow can be reduced by 10 percent by throttling the discharge valves associated with some users without adversely affecting production or increasing maintenance requirements. From the field data, it is estimated that the operating discharge pressure will be controlled at 40 psig (2.8 bar). Electrical operating costs associated with Scenario II are summarized in Table 14.2.

Typical installed costs over and above those necessary for the installation described in Scenario I are tabulated below.

Item	Installed Cost
Variable Frequency Drive	15,000
Pressure Transmitter	1,000
Controller	2,000
TOTAL	18,000

It is assumed that the existing combination starter will eventually be used elsewhere in the plant, avoiding the expenditure of approximately US $4000; however, this has not been used in the calculations.

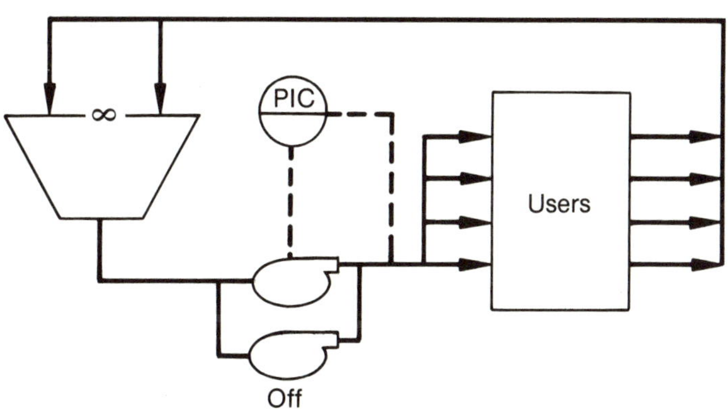

Figure 14.7
Scenario II — Circulating Water System (Proposed)

Table 14.2
Scenario II
Circulating Water System

Flow[1] (gpm)	Discharge Pressure[2] (psig)	Motor Speed[3] (%)	Motor Capacity[4] (hp)	Actual Motor Load[5] (bhp)	Power Consumption[6] (kW)	Operating Time[7] (%)	Weighted Power Consumption[8] (kW)
0	0	0.0	0.0	0.0	0.0	5	0.0
1260	40	74.8	74.8	34.7	34.7	15	5.2
1350	40	75.1	75.1	35.1	35.1	50	17.5
1440	40	75.4	75.4	35.6	35.4	15	5.3
1530	40	75.8	75.8	36.1	35.9	10	3.6
1620	40	76.3	76.3	36.8	36.4	5	1.8
					Total	100	33.5
					Annual Cost/kW		539.36
					Annual Operating Cost		18046

1. Estimated at 90% of Scenario I flow.
2. Controlled.
3. See Chapter 13 for equations or pump curve;
   Motor Capacity = 100 hp, $\Delta\eta$ = 0.03 (at all speeds), Design Pressure = 64.9 psig, Design Flow = 1800 gpm, Deadhead Pressure = 77.1 psig.
4. Motor Speed × Full Speed Motor Capacity/100.
5. Equation 13.2, bhp at Design Conditions = 83 hp.
6. $V = I \times R$.
7. Estimated.
8. Power Consumption × Operating Time/100.

Circulating Water System Analysis

The scenarios presented above can be summarized on the following table for analysis.

Scenario	Installed Cost	Operating Cost	Savings
I	US$00 0	US$34,718	US$ 0
II	US$18,000	US$18,046	US$16,672

From the above, it can be seen that Scenario II offers a simple payback of 18,000/(34,718 – 18,046), or 1.1 years. The percentage of electrical savings is (64.4 kW – 33.5 kW)/64.4 kW, or 48.0 percent, which illustrates the potential of this technology to reduce electrical energy consumption. From the above analysis, it can be seen that application represents a significant economic opportunity, due to the substantial percentage reduction of electrical energy consumption of a relatively large motor.

CENTRIFUGAL FAN SYSTEMS

Centrifugal fans are used throughout industry and in buildings, especially in heating, ventilating and air conditioning (HVAC) applications, for the movement of gases (typically air). In most applications, centrifugal fan systems have low static head requirements and, as a result, significant speed reduction can often be achieved, which results in corresponding significant electrical energy savings.

Scenario I

A 30 horsepower (22.4 kW) motor operating at 1800 rpm with a brake horsepower of 27 hp (20.1 kW) to be installed to drive a centrifugal fan and provide 20,000 cfm (566 m^3/min) of air flow at a discharge pressure of 6 inches (152 mm) WC, as illustrated on the fan curve shown in Figure 14.8. Outlet fan damper control is proposed for the new installation as illustrated in Figure 14.9.

As centrifugal fan curves are similar in nature to centrifugal pump curves and the damper is analogous to a control valve on the pump discharge, the calculation of the operating cost of this system is similar to that of a centrifugal pump and is summarized in Table 14.3.

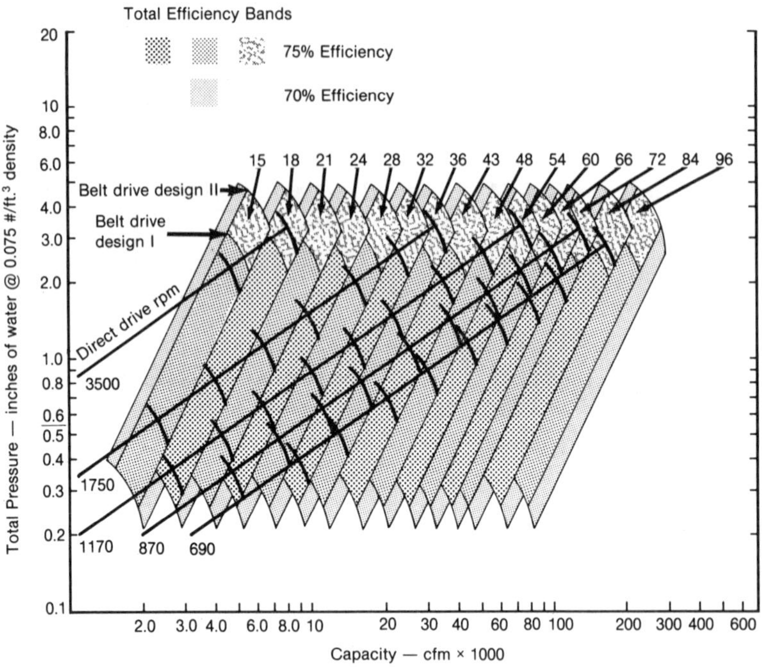

Figure 14.8
Fan Curve

(courtesy Buffalo Forge)

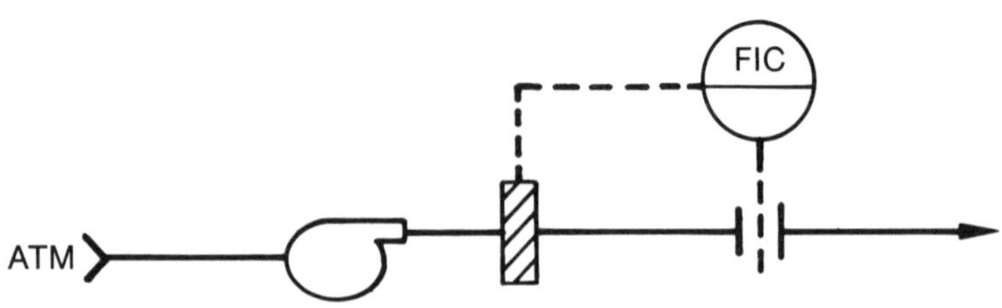

Figure 14.9
Scenario I — Centrifugal Fan System (Proposed Outlet Damper Control)

Table 14.3
Scenario I
Centrifugal Fan System

Flow[1] (cfm)	Actual Motor Load[2] (bhp)	Power Consumption[3] (kW)	Operating Time[1] (%)	Weighted Power Consumption[4] (kW)
0	0.0	0.0	5	0.0
12,000	23.5	22.3	15	3.3
14,000	24.5	22.9	50	11.5
16,000	25.5	23.6	15	3.5
18,000	26.5	24.3	10	2.4
20,000	27.0	24.6	5	1.2
		Total	100	22.0
		Annual Cost/kW		539.36
		Annual Operating Cost		11868

1. Estimated.
2. From pump curve.
3. $V = I \times R$, Motor Capacity = 30 hp, $\Delta\eta = 0.09$.
4. Power Consumption × Operating Time/100.

Scenario II

As shown in Figure 14.10, Scenario II entails locating the fan damper at the fan suction, reducing the suction pressure, and reducing the gas density and electrical energy requirements by an estimated 10 percent. As a result, the operating costs associated with Scenario II are 0.90 × US$11,868, or US$10,681. Installed cost is identical to that of Scenario I.

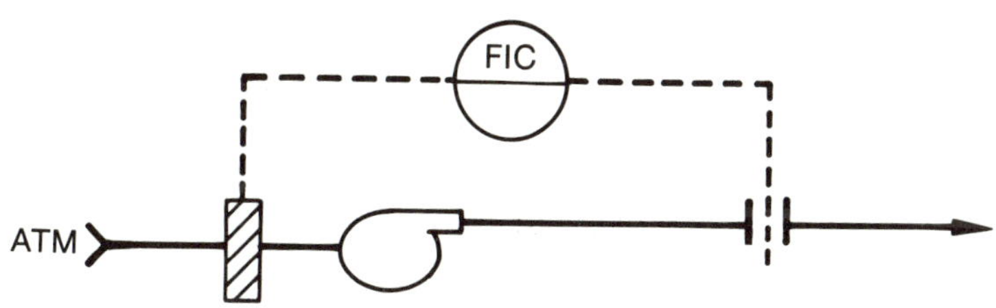

Figure 14.10
Scenario II — Centrifugal Fan System (Inlet Damper Control)

Scenario III

Scenario III entails varying fan speed with an electronic variable frequency drive to control flow, as illustrated in Figure 14.11.

The electrical energy cost associated with this scenario is summarized in Table 14.4.

Typical installed costs over and above those necessary for the installation described in Scenario I are tabulated below.

Item	Installed Cost
Variable Frequency Drive	5,000
Fan Damper and Actuator	(4,000)
Combination Starter	(1,500)
TOTAL	(500)

Note that this installation is less expensive than that of Scenario I.

Scenario IV

One scenario often suggested, because of its conceptual and operational simplicity, is control of the desired flow by bleeding off excess capacity, as shown in Figure 14.12.

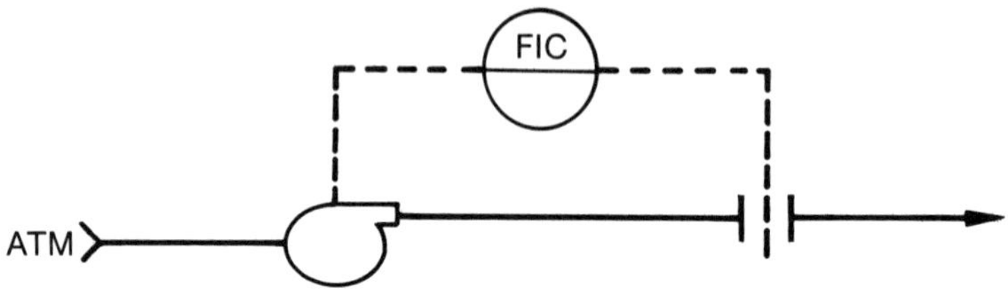

Figure 14.11
Scenario III — Centrifugal Fan System (Variable Frequency Drive)

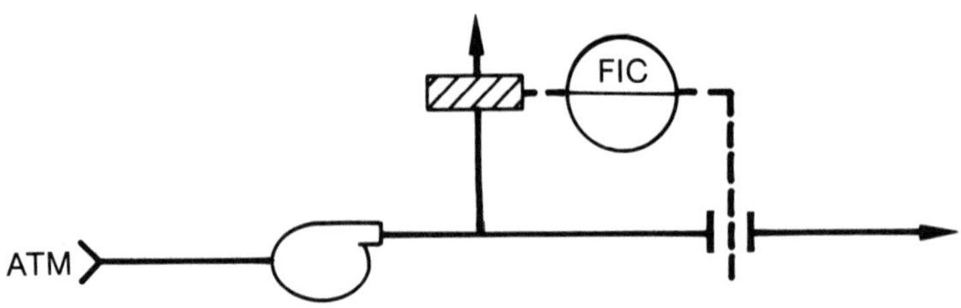

Figure 14.12
Scenario IV — Centrifugal Fan System (Bleed-off Control)

Table 14.4
Scenario III
Centrifugal Fan System

Flow[1] (cfm)	Discharge Pressure[2] (″ WC)	Motor Speed[3] (%)	Motor Capacity[4] (hp)	Actual Motor Load[5] (bhp)	Power Consumption[6] (kW)	Operating Time[7] (%)	Weighted Power Consumption[8] (kW)
0	0	0.0	0.0	0.0	0.0	5	0.0
12,000	2.9	63.8	19.1	7.0	8.8	15	1.3
14,000	3.2	72.2	21.7	10.2	11.5	50	5.7
16,000	3.6	77.8	23.3	12.7	13.6	15	2.0
18,000	4.0	86.1	25.8	17.2	17.1	10	1.7
20,000	4.4	91.7	27.5	20.8	19.9	5	1.0
					Total	100	11.8
					Annual Cost/kW		539.36
					Annual Operating Cost		6366

1. Estimated.
2. See Chapter 13 for equations.
   $F_{XS\ cap}$ = 15%, Fcv = 30%, Design Pressure = 6″ WC, Static Pressure = 2″ WC.
3. Fan Curve.
   Motor Capacity = 30 hp, $\Delta\eta$ = 0.09 (at all speeds), Design Flow = 20,000 cfm, Deadhead Pressure = 8.4″ WC.
4. Motor Speed × Full Speed Motor Capacity/100.
5. bhp at Design Conditions = 27 hp.
6. $V = I \times R$.
7. Estimated.
8. Power Consumption × Operating Time/100.

In this scenario, the fan operates continually at full capacity regardless of load conditions. The electrical energy consumption associated with this scenario is approximately 24.6 kW × US$539.36/kW·yr, or US$13,268. The installed cost is identical to that of Scenario I.

Analysis

The scenarios presented above can be summarized on the following table for analysis.

Scenario	Installed Cost	Operating Cost	Savings
I	US$ 0	US$11,868	US$ 0
II	US$ 0	US$10,088	US$ 1,780
III	US$(500)	US$ 6,366	US$ 5,502
IV	US$ 0	US$13,268	US$(1,400)

This table shows that the installation cost of all of the scenarios can be considered equal. In comparison to Scenario I, Scenario II reduces operating costs slightly, while Scenario IV increases operating costs. Scenario III offers operational cost savings representing a reduction of electrical energy consumption of (22.0 – 11.8)/22.0, or 46.4 percent.

AIR COMPRESSOR SYSTEM

Positive displacement air compression systems are used extensively in industry and in commercial buildings as the motive force for valves, dampers, tools, and the like. Various types of air compressors exist, giving rise to many different control strategies.

Scenario I

The existing 2000 cubic feet per minute (56.6 m^3/min) air compressor with a 500 horsepower (373 kW) motor (see Figure 14.13) is loaded until a high pressure set point is reached, at which time the compressor unloads until the low pressure set point is reached and the compressor loads.

On a large existing motor, it is preferred that operating cost calculations be based on actual measurements, which are summarized below.

Time loaded 30 seconds @ 400 kW
Time unloaded 30 seconds @ 60 kW

$$\text{Average kW} = \frac{(400 \text{ kW} \times 30 \text{ sec} + 60 \text{ kW} \times 30 \text{ sec})}{60 \text{ sec}} = 230 \text{ kW}$$

The operating cost is 230 kW × US$539.36/kW·yr, or US$124,052 per year.

Scenario II

Varying compressor speed to match load conditions will minimize the time during which the compressor is unloaded and not performing any work as illustrated in Figure 14.14. To verify that the compressor speed can be reduced without adverse effects, the compressor manufacturer was consulted and stated that the compressor must operate between 50 and 100 percent of speed to provide sufficient lubrication and cooling to avoid equipment damage. The variable speed drive will be adjusted so that it cannot operate below 50% speed.

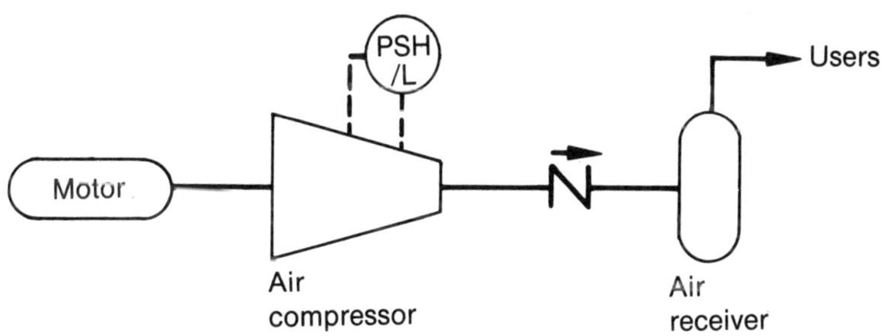

Figure 14.13
Scenario I — Air Compressor System (Existing Load/Unload Control)

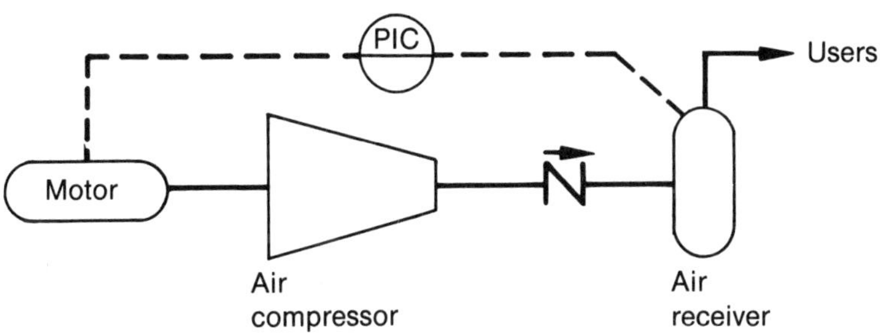

Figure 14.14
Scenario II — Air Compressor System (Variable Frequency Drive)

The air compressor operates at 30/(30 + 30), or 50 percent of capacity (1000 cfm) and 50 percent speed, assuming no appreciable drop in efficiency. The operating cost associated with the above system is 400 kW × 50 percent × US$539.36, or US$107,872.

Typical installed costs over and above those necessary for the installation described in Scenario I are tabulated below.

Item	Installed Cost
Variable Frequency Drive	40,000
Pressure Transmitter	1,200
Controller	2,000
TOTAL	43,200

Scenario III

In applications where the air compressor is located near a liquid fuel burning boiler, hot compressed air can replace atomizing steam and reduce overall operating costs. The hot air can be taken directly off the compressor discharge, but should the compressor unload, the boiler will shut down. Therefore, this scenario includes the equipment in Scenario II, hot air piping, valves and interlocks.

This application (see Figure 14.15) differs from the previous scenarios in that the economic ramifications of the process change must be calculated. The following process data was obtained from the Utility Engineer in the boiler house.

Boiler Fuel	Oil
Fuel Oil Density	8.1 lb/gal
Fuel Oil Energy Content	150,000 Btu/gal
Fuel Cost	US$3.40/1,000,000 Btu
Average Steam Load	75,000 lb/hr
Energy Input	1200 Btu/lb of steam
Boiler Efficiency	75 percent
Stack Temperature	450°F
Steam Temperature	350°F
Hot Air Temperature	300°F
Atomization Steam	1/3 lb/lb fuel

The amount of oil consumed per hour can be calculated as

$$\text{Fuel consumption} = \frac{(75{,}000 \text{ lb steam/hr})\ (1200 \text{ Btu/lb steam})}{(0.75 \text{ Boiler efficiency})}$$

$$\times \frac{(8.1 \text{ lb oil/gal})}{(150{,}000 \text{ Btu/gal oil})} = 6480 \text{ lb/hr oil}$$

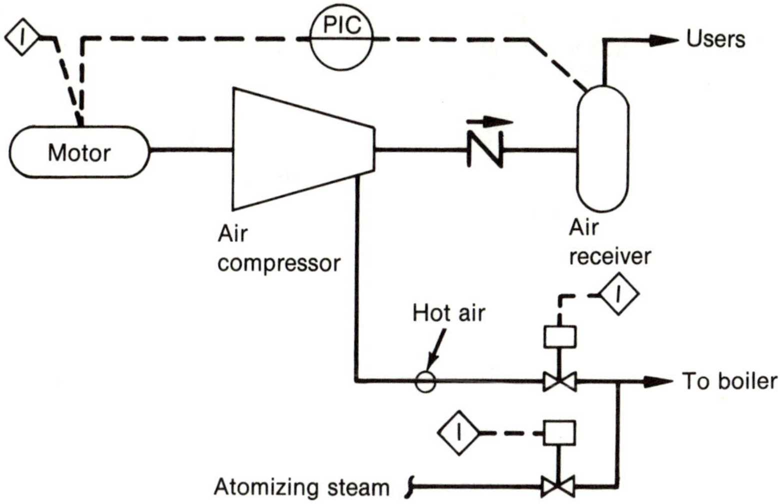

Figure 14.15
Scenario III — Air Compressor System
(Variable Frequency Drive with Hot Air Atomization)

while the amount of atomizing gas can be estimated by assuming that the mass of atomizing gas is 1/3 that of the mass of the liquid fuel.

$$\text{Atomizing gas} = \left(\frac{1}{3}\ \frac{\text{lb atomizing gas}}{\text{lb oil}}\right)(6480\ \text{lb/hr oil})$$

$$= 2160\ \text{lb/hr atomizing gas}$$

Due to the substitution of one atomizing gas for another, the cost of each gas is calculated to determine the economic effects of the modification. The hourly fuel cost associated with the generation of atomizing steam can be calculated as

$$\text{Steam cost} = (2160\ \text{lb/hr})\left(\frac{1200\ \text{Btu}}{\text{lb}}\right)\left(\frac{\text{US\$3.40}}{1{,}000{,}000\ \text{Btu}}\right)\left(\frac{1}{0.75\ \text{efficiency}}\right)$$

$$= \text{US\$11.75/hr}$$

while the hourly cost of hot air production is

$$\text{Hot air cost} = \frac{(2160\ \text{lb/hr})}{(0.075\ \text{lb/ft}^3)}\left(\frac{400\ \text{kW}}{2000\ \text{ft}^3/\text{min}\cdot 60\ \text{min}}\right)\left(\frac{\text{US\$539.36/kW}\cdot\text{yr}}{8760\ \text{hr/yr}}\right)$$

$$= \text{US\$5.91/hr}$$

As hot air is less expensive to produce than steam, a net savings of (US\$11.75 – US\$5.91) (8760 hrs/yr), or US\$51,158 per year results from the substitution of hot air for steam.

As hot air and steam have different temperatures and specific heats, the amount of heat that each gas carries up the stack is different and must be calculated. For steam atomization, the operating cost of the losses is calculated as

$$\begin{aligned}\text{Stack loss (steam)} &= mC_p\,\Delta T\\ &= (2160\ \text{lb/hr})\ (0.50\ \text{Btu/lb}\cdot{}^\circ\text{F})\\ &\qquad (450 - 350^\circ\text{F})\\ &= 108{,}000\ \text{Btu/hr}\end{aligned}$$

$$\text{Cost} = (108{,}000\ \text{Btu/hr})\left(\frac{\text{US\$3.40}}{1{,}000{,}000\ \text{Btu}}\right)(8760\ \text{hrs/yr}) = \text{US\$3217/yr}$$

For hot air atomization, the costs are

$$\begin{aligned}\text{Stack loss (air)} &= (2160\ \text{lb/hr})\ (0.25\ \text{Btu/lb}\cdot{}^\circ\text{F})\ (450 - 300^\circ\text{F})\\ &= 80{,}000\ \text{Btu/hr}\end{aligned}$$

Fuel savings of (US\$3217 – US\$2383) or US\$834 per year result from the substitution of hot air for atomizing steam.

The additional air requirements associated with this scenario can be calculated as

$$\left[\frac{(2160\ \text{lb/hr})}{(0.075\ \text{lb/ft}^3)\ (60\ \text{min/hr})}\right] \text{or 480 cfm,}$$

which is within the capacity of the compressor.

Electrical costs can be calculated as

$$\left(\frac{1000 + 480\ \text{cfm}}{2000\ \text{cfm}}\right)(400\ \text{kW})\ (\text{US\$539.36/kW}\cdot\text{yr})\ \text{or US\$159,651.}$$

The total operating cost is US\$159,651 + US\$2383, or US\$162,976 per year.

To properly compare the various scenarios, the operating costs associated with steam atomization — (US\$11.75/hr) (8760/yr) + US\$3217 or US\$106,147 — must be added to Scenarios I and II. The operational costs are then (US\$124,052 + US\$106,104) or US\$230,156 and (US\$107,872 + US\$106,104) or US\$213,976 for Scenarios I and II, respectively.

Typical installed costs over and above those necessary for the installation described in Scenario I are tabulated below.

Item	Installed Cost
Variable Frequency Drive	40,000
Pressure Transmitter	1,200
Controller	2,000
Hot Air Piping and Valves	10,000
TOTAL	53,200

Scenario IV

A constant air supply can be obtained by controlling receiver pressure by discharging excess capacity to atmosphere (see Figure 14.16).

With this control strategy, capital expenditures are low, but the compressor remains loaded regardless of actual load and the operating load is

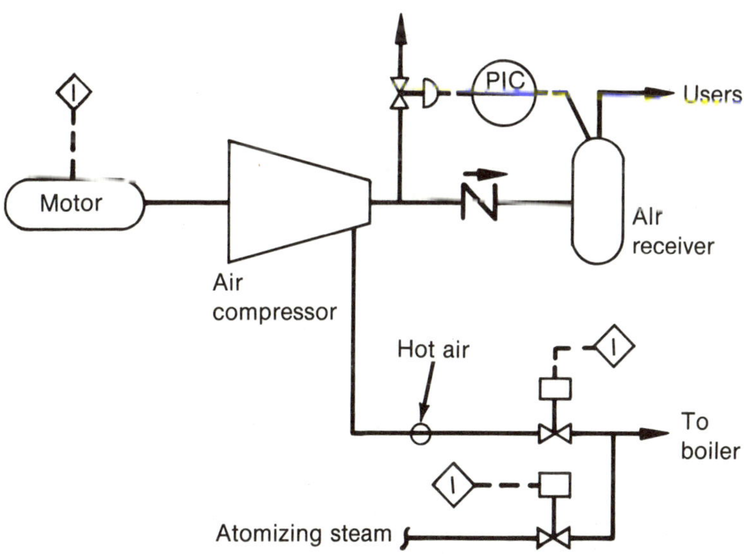

Figure 14.16
Scenario IV — Air Compressor System (Bleed-off Control)

400 kW. The total operating cost associated with this scenario is the electrical operating costs (400 kW × US$539.36/kW·yr = US$215,744/yr) plus the cost of atomization stack losses (US$2383) or US$218,127.

Typical installed costs over and above those necessary for the installation described in Scenario I are tabulated below.

Item	Installed Cost
Pressure Transmitter	1,200
Controller	2,000
Control Valve	3,000
Hot Air Piping and Valves	10,000
TOTAL	16,200

Analysis

The scenarios presented above can be summarized on the following table for analysis.

Scenario	Installed Cost	Operating Cost	Savings	Simple Payback
I	US$ 0	US$230,156	US$ 0	—
II	US$43,200	US$213,976	US$16,180	2.7 yr
III	US$53,200	US$162,976	US$67,182	0.8 yr
IV	US$16,200	US$218,127	US$12,029	1.3 yr

From the analysis, it can be seen that while the installation of a variable speed drive may be conceptually superior, the economic justification may be marginal (Scenario II). However, variable speed drive technology coupled with process improvement can yield very attractive economic justification (Scenario III). Pitfalls, such as the large increase in electrical consumption in return for small operational savings (Scenario IV), should be avoided. Note that the payback associated with designs such as those illustrated by Scenario IV can be attractive, due to the low amount of capital investment required for implementation. However, large potential savings can remain after such an "economical" design is implemented.

LARGE FAN SYSTEM

While electronic variable frequency drives are applicable to the great majority of instances, there are many applications, such as the case of a large fan, where utilization of another variable speed drive technology is desirable.

Scenario I

The proposed flow control strategy of a 10,000 horsepower (7460 kW) centrifugal fan is to vary flow by varying the position of the inlet damper, as illustrated in Figure 14.17.

Assuming that the fan is operated 50 percent of the time at minimum flow (1000 bhp) and 50 percent of the time at design conditions (9500 bhp), the electrical operating costs associated with this control strategy can be summarized as shown in Table 14.5.

Table 14.5
Scenario I
Large Centrifugal Fan System

Flow[1] (%)	Actual Motor Load[2] (bhp)	Power Consumption[3] (kW)	Operating Time[1] (%)	Weighted Power Consumption[4] (kW)
0	1000.0	1885.1	50	942.5
100	9500.0	7680.6	50	3840.3
		Total	100	4782.8
		Annual Cost/kW		539.36
		Annual Operating Cost		2579663

1. Estimated.
2. From fan curve for inlet damper control.
3. Motor Capacity = 10,000 hp, $\Delta\eta = 0$.
4. Power Consumption × Operating Time/100.

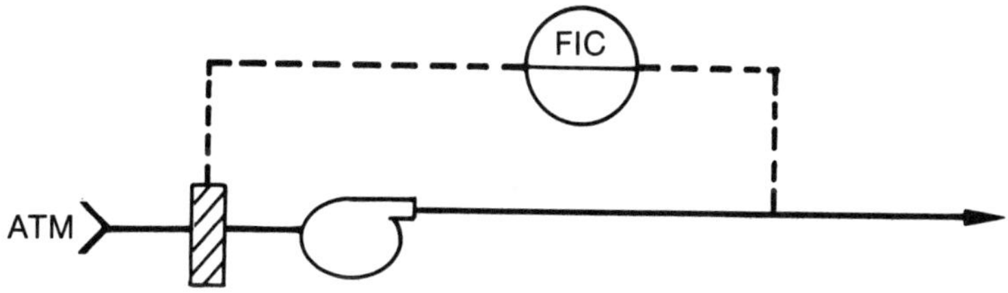

Figure 14.17
Scenario I — Large Centrifugal Fan System (Proposed)

Scenario II

Due to the size of the motor, a hydraulic variable speed drive is proposed to control flow by varying the speed of the fan while the motor continues to operate at full speed (see Figure 14.18).

Noting that the motor operates at full speed, the electrical operating costs associated with this scenario are summarized in Table 14.6.

Typical installed costs over and above those necessary for the installation described in Scenario I are tabulated below.

Item	Installed Cost
Hydraulic Variable Speed Drive	600,000
Damper	(10,000)
TOTAL	590,000

The installation of the above equipment will increase mechanical maintenance costs, but the amount can not be readily quantified.

Analysis

The scenarios presented above can be summarized on the following table for analysis.

Scenario	Installed Cost	Operating Cost	Savings
I	US$ 0	US$2,579,663	US$ 0
II	US$590,000	US$1,102,610	US$1,477,053

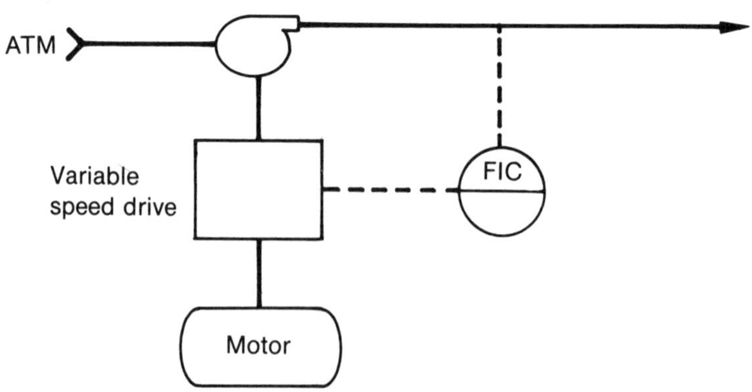

Figure 14.18
Scenario II — Large Centrifugal Fan System (Variable Speed Drive)

Table 14.6
Scenario II
Large Centrifugal Fan System

Flow[1] (%)	Discharge Pressure[2] (″ WC)	Fan Speed[3] (%)	Motor Capacity[4] (hp)	Actual Motor Load[5] (bhp)	Power Consumption (kW)	Operating Time[6] (%)	Weighted Power Consumption[7] (kW)
0	0	0.0	10,000.0	0.0	1203.2	50	601.6
100	4.4	63.8	10,000.0	2467.1	2885.4	50	1442.7
					Total	100	2044.3
					Annual Cost/kW		539.36
					Annual Operating Cost		1102610

1. Estimated.
2. Chapter 13 equations; $F_{XS\ cap}$ = 15%, Fcv = 30%, Design Pressure = 6″ WC, Static Pressure = 2″ WC, ″ WC. C.
3. Fan Curve.
   Motor Capacity = 10,000 hp, $\Delta\eta$ = 0 (at all speeds), Design Flow = 100%, Deadhead Pressure = 8.4″ WC.
4. Full Speed Motor Capacity.
5. bhp at Design Conditions = 9500 hp.
6. Estimated.
7. Power Consumption × Operating Time ' 100.

From the above, it can be seen that Scenario II significantly reduces operating costs and offers a simple payback of 590,000/(2,579,663 – 1,102,610), or 0.40 years.

SUMMARY

The above scenarios represent but a few of the many viable applications of variable speed drive technology that have merit based on a combination of electrical and process cost savings. Many applications are similar to those described above, where the the operating times, excessive power factor penalties, hatchet/ratchet clauses, and the like, should be taken into account. Therefore, it is important to understand that the examples were developed to illustrate the concepts of justifying variable speed drive technology applications rather than as definitive case studies.

In many applications, it may be difficult to justify retrofitting the existing motors and controls with variable speed drives. Nevertheless, each motor should be viewed as a potential application, as many economically attractive retrofit applications do exist. Due to the relatively low installed costs associated with applications, every motor should be considered a likely candidate for a variable speed drive until proven otherwise. It is estimated that over 50 percent of all motors in new continuous processes should have variable speed drive control due to (1) a lower installed cost than an equivalent loop with a control valve, (2) a reduction in operating costs, or (3) an improvement of process control.

Properly applied, variable speed drive technology can represent seizure of technical and economic opportunity that can yield handsome rewards.

Bibliography

Chemical Rubber Co. *Handbook of Physica and Chemistry.* Cleveland, OH, 1971.

Crane Company. "Flow of Liquids through Valves, Fittings and Pipe," Technical Paper No. 410. 17th printing, 1978.

Driskell, Les. *Introduction to Control Valves and Other Final Control Elements.* Research Triangle Park, NC: Instrument Society of America, 1981.

Dukelow, Sam G. *The Control of Boilers.* Research Triangle Park, NC: Instrument Society of America, 1986.

Editors. "1987 Control Products Specifier." *Control Engineering.* Nov. 1986.

Gillum, Donald R. *Industrial Pressure Measurement.* Research Triangle Park, NC: Instrument Society of America, 1982.

Gillum, Donald R. *Industrial Level Measurement.* Research Triangle Park, NC: Instrument Society of America, 1984.

Goulds Pumps, Inc. *Gould's Pump Manual,* "Technical Data" section. Seneca Falls, NY.

Hutchison, J. W. *ISA Handbook of Control Valves.* 2nd ed. Research Triangle Park, NC: Instrument Society of America, 1976.

Instrument Society of America. *ISA Directory of Instrumentation.* Research Triangle Park, NC: 1986.

Kerlin, Thomas W. and Shepherd, Robert L. *Industrial Temperature Measurement.* Research Triangle Park, NC: Instrument Society of America, 1982.

Lange. *Handbook of Chemistry.* 12th ed. NY: McGraw-Hill Book Co., 1979.

Mollenkamp, Robert A. *Introduction to Automatic Process Control.* Research Triangle Park, NC: Instrument Society of America, 1984.

Murrill, Paul W. *Fundamentals of Process Control Theory.* Research Triangle Park, NC: Instrument Society of America, 1981.

North American Manufacturing Co. *Combustion Handbook.* 2nd ed. Cleveland, OH: 1978.

Perry, Robert H. and Chilton, Cecil H. *Chemical Engineer's Handbook.* 5th ed. NY: McGraw-Hill Book Co., 1973.

Sears, Francis W. and Zemansky, Mark W. *University Physics.* 2nd ed. NY: Addison-Wesley Publishing Co., Inc., 1955.

Spitzer, David W. *Industrial Flow Measurement.* Research Triangle Park, NC: Instrument Society of America, 1984.